Zero to Genetic Engineering Hero

*The beginner's guide to programming bacteria
at home, school & in the makerspace*

First Edition

Justin Pahara, Julie Legault

Make: Zero to Genetic Engineering Hero

By Justin Pahara and Julie Legault

Published by Make: Community LLC
150 Todd Road, Suite 200, Santa Rosa, CA 95407
Make: books may be purchased for educational, business, or sales promotional use. Online editions are also available for most titles.

Publisher: Dale Dougherty
Illustrations: Julie Legault
Cover Design: Jason Babler

January 2021: First Edition
September 2021: Second Edition

See www.oreilly.com/catalog/errata.sp?isbn=97816804 57162 for release details.

HOW TO CONTACT US:
Please address comments and questions concerning this book to the publisher via postal mail, or via email at books@make.co.

Make: Community is a growing, global association of makers who are shaping the future of education and democratizing innovation. Through *Make:* magazine, and 200+ annual Maker Faires, *Make:* books, and more, we share the know-how of makers and promote the practice of making in schools, libraries and homes. To learn more about *Make:* visit us at make.co.

IMPORTANT MESSAGE TO OUR READERS:
Your safety is your responsibility, including proper use of equipment and safety gear. Chemicals, ingredients, biochemicals, electronics and other resources used for these experiments are dangerous unless used properly and with adequate precautions, including safety gear. Safety precautions outlined in this book, including guidelines on how to set up your experiment space, must be followed. Some illustrations and photos do not depict safety precautions or equipment in order to show the experiment steps more clearly. These experiments are not intended for use by anyone under 12 years old. Adult supervision is recommended.

Use of the instructions and suggestions in *Zero to Genetic Engineering Hero* is at your own risk. The publishers, the authors, and the technology manufacturers disclaim all responsibility for any resulting damage, injury, or expense. It is your responsibility to make sure that your activities comply with applicable laws, including copyright.

978-1-6804-5716-2

Contents

Authors **6**

Junior Editors **7**

Preface **8**
    Who this book is for: 9
    This book is not a typical textbook! 9
    Acknowledgements 9

The Genetic Engineers' Pledge **11**

Isolating DNA, the Blueprints of Life **12**

    Equipment and Materials **13**
    Learning Hands-On: Breaking Cells Open & Extracting DNA 15
    Step 1. Create salt water inside the resealable bag 15
    Step 2. Mashing the strawberry into individual cells 16
    Step 3. Breaking open the cells with soap 17
    Step 4. Filtering the cell debris 19
    Step 5. Precipitating the DNA 19

    Fundamentals: DNA **23**
    Evolution: It's natural for DNA to change... 23
    Modern Synthesis.. 24
    Genetic engineering: The road to precise editing of DNA 26
    Atoms, molecules, and macromolecules of the cell............................... 27
    Understanding the nomenclature of DNA 32
    DNA extractions in the real-world 34

    Summary and What's Next? **36**

Setting Up Your Genetic Engineering Hero Space **39**

    Getting Started **40**
    Do I need government approval? 40
    What type of room should I set up my genetic engineering space in?.... 41
    Equipment and materials for your Genetic Engineering Hero space....... 42
    Materials and Supplies 44
    Cleaning and Other Supplies 45
    Experiment kits/wetware 46
    Safety and Best Practices 46
    Just because I Can, does it mean I should?... 47

Growing *E. coli* Cells **51**

    Getting Started **53**
    Equipment and Materials.. 53
    Virtual Bioengineer™ simulation Breakout Session 2 54
    Learning Hands-On: Growing K12 *E. coli* cells 55
    Step 1. Download the instruction manual for the Canvas Kit 55
    Step 2. Put on your gloves and lab coat 55
    Step 3. Create molten LB agar powder 56
    Step 4. Adding antibiotic 57
    Step 5. Pour LB agar plates 58
    Step 6. Use & storage of LB agar plates 60
    Step 7. Streaking *E. coli* bacteria 61
    Step 8. Incubating *E. coli* cells 61
    Step 9. Viewing plates of grown bacteria 62
    Step 10. Painting living art (bioart) with *E. coli* bacteria 63
    Step 11. Incubating *E. coli* cells 64
    Step 12. Viewing & Preserving your bioart with a Keep-it Kit 64
    Step 13. Clean-up and inactivation 65

Fundamentals: *E. coli* Cells **67**
Introduction to "Lab" *E. coli*..**67**
A Tour of the *E. coli* Microfactory 70
The Fence (A)... 70
The Outer Wall (B)... 72
The Lobby (C) 74
The Inner Walls (D)... 75
The Factory Floor (E).. 75

Summary and What's Next? **83**

Genetic Engineering Your *E. coli* Cells 8 6

Getting Started **87**
Equipment and Materials.. 87
Learning Hands-On: Transforming K12 *E. coli* cells with a DNA Plasmid 89
Step 1. Download the instruction manual for the Engineer-it Kit 89
Step 2. Put on your gloves and lab coat 89
Step 3. Label your plates 90
Step 4. Make non-selective and selective LB agar plates 90
Step 5. Streaking *E. coli* and the negative control plates 92
 Check Point! 94
Step 6. Making chemically competent cells 95
Step 7. Add DNA plasmids and Heat Shock 96
Step 8. Recovery step 98
Step 9. Plating and incubating your cells 99
Step 10. What to expect & inactivation 100

Fundamentals: How a cell reads a DNA plasmid **101**
The basic operating environment of a cell: The *Four B's* 101
(Bump, Bind, Burst, Bump) 101
The *Three Steps to Microfacturing* 102
Deoxyribonucleic acid (DNA) vs. Ribonucleic acid (RNA).. 103
RNA polymerase: The cell machine that transcribes.. 105
What is a gene? 106
Starting Transcription.. 106
During transcription: Direction 109
During transcription: Which DNA strand does RNA polymerase read?...................... 110
During transcription: A secret cipher for transcribing DNA to RNA 112
Stopping transcription 113
What can you do with RNA?.. 114

Summary and What's Next? **116**

Extracting your engineered proteins 119

Getting Started **120**
Equipment and Materials.. 120
Learning Hands-On: Culture and lyse engineered *E. coli* to obtain a protein product extract 121
Step 1. Download the instruction manual for the Plate Extract-it Kit 121
Step 2. Put on your gloves and lab coat 121
Step 3. Transform cells to get fresh colonies (Optional) 121
Step 4. Make selective LB agar plates for amplification 122
Step 5. Culturing: Spread out your freshly engineered cells 122
Step 6. Culturing: Incubate at 37 ˚C for 24-48 hours 123
Step 7. Extraction: Collect cells and start the lysis 125
Step 8. Extraction: Lyse the cells 126
Step 9.Extraction: Pellet the cell debris 127
Step 10. Extraction: Filter sterilize your proteins 128
Step 11. Using your proteins 129

Fundamentals: How cells translate proteins from RNA **131**
Step two of the *Three Steps to Microfacturing*: Translating proteins from RNA 131
Starting Translation.. 132
During Translation: The RNA to protein cipher... 133
During Translation: Locating the starting point for translation.................................... 137
Stopping Translation 139

Summary and What's Next? 142
    Check Point! 145

Processing Enzymes 147

Getting Started 149
Equipment and Materials... 149
Learning Hands-On: Process one molecule into another using enzymes 150
Exercise 1: Enzymatic processing to generate smells 150
    Step 1. Download the instruction manual for the Smell-it Kit 150
    Step 2. Complete the engineering part of the Smell-it Kit 150
    Step 3. Culture cells with the substrate 150
    Step 4. Labeling and creating your LB agar plates for culturing 151
    Step 5. Culture engineered cells on your selective LB agar plate 152
Exercise 2: Enzymatic processing to generate color 153
    Step 1. Download the instruction manual for the Blue-it Kit 154
    Step 2. Complete genetic engineering and extraction procedures 154
    Step 3. Dissolve your substrates 154
    Step 4. Add cell extract beta-galactosidase to the substrate 155

Fundamentals: Diving into enzymatic processing 159
The basics of enzymatic chemical reactions... 159
The *Four B's* and enzyme function 159
Atoms 161
Bonds 163
Protein enzyme catalysis in cells 168

Summary and What's Next? 174

Manually turning on genes *in situ* 177

Getting Started 178
Equipment and Materials... 178
Learning Hands-On: Manually turning on genes in situ 179
Exercise 1: Inducing a gene using a chemical 179
    Step 1. Complete the Induce-it Kit engineering exercise. 179
    Step 2. Culture your cells 179
    Step 3. Add your inducer 179
Exercise 2: Inducing a gene using temperature 180
    Step 1. Complete the Heat-it Kit engineering 180
    Step 2. Increase the temperature 181
Exercise 3: Inducing a gene using light. 181
    Step 1. Streak cells 181
    Step 2. Turn on the light! 181

Fundamentals: Diving deeper into genetic 'switches' 183
Turn on genes with chemicals... 183
Turn on genes with temperature 185
Turn on genes with light... 186

Summary 188

Periodic table 192

Authors

@jpahara

Dr. Justin Pahara is a Cree scientist-entrepreneur from a Southern Alberta Canadian farm. Over the last decade, Justin has worked in pure and applied life sciences inside academia and out. Justin began his career earning a B. Sc. (Immunology and Infection), and an M. Sc. (Cell Biology) from the University of Alberta. Justin furthered his skills and knowledge by completing a Ph.D. (Biotechnology & Bioelectronics) at the University of Cambridge in the United Kingdom. Since being an early participant in the International Genetically Engineered Machines (iGEM) competition in 2007, Justin has continued the pursuit of understanding and implementing genetic engineering and synthetic biology in his academic and entrepreneurial career.

During his doctorate, Justin attended the Graduate Studies Program (GSP) at Singularity University, a professional development program in the heart of Silicon Valley that focuses on exponential technology. Justin's entrepreneurial passion was ignited by the high caliber faculty, peers and rich entrepreneurship content.

Following the program, and after completing his doctorate, Justin entered the world of entrepreneurship, first with a DNA technology and IT startup, Synbiota. In the pursuit of "exponentially increasing global bio-innovation", Justin joined Amino Labs in 2016, an MIT Media Lab spinout that builds hardware and synthetic biology products making genetic engineering accessible to children and non-scientists.

Throughout his career, Justin participated and mentored in several entrepreneurship environments, including RebelBio, IndieBio (Biotechnology Startup Accelerators), Mozilla WebFWD (OpenSource Software Startup Accelerator) and helped start several Do-It-Yourself Bio labs in Canada. Justin is an *Emerging Leader of Biosecurity Initiative* Fellow.

julielegault.com

Julie Legault is a designer-entrepreneur from the city of Montreal. Over the last decade, Julie has worked in design research, innovation and the maker movement. Julie began her career earning a B.F.A. (Design & Computation Art) and a Graduate Certificate (Digital Technologies in Design Art) from Concordia University in Montreal. Julie furthered her skills and knowledge by completing a Master of Art in the School of Materials at the Royal College of Art in London, United Kingdom and a Master of Science at the Massachusetts Institute of Technology (MIT) Media Lab. Grateful for her early access to computing in her childhood thanks to a pioneering mother, Julie has since dedicated herself to translating complex technologies for beginners through teaching and applied design.

Throughout, Julie participated in design residencies worldwide, taught at Birmingham's Institute of Art and Design, worked with multi-nationals, and pop stars to develop accessible smart materials, wearables and biometric devices. Her work has been published and exhibited globally, notably in Wired Magazine, NPR, MoMA, NYT, ARS Electronica and the V&A.

As an adult, her unexpected foray into DNA technology impacted her so thoroughly that finding an accessible entry point in the complex science became the focus of her MIT Media Lab graduate thesis. This led her to found Amino Labs, an MIT spinout that designs and builds equipment and experiences to make genetic engineering and biotechnology accessible to youth and non-scientists.

Since then, Julie has participated and mentored in several entrepreneurship environments, including IndieBio (Silicon Valley Biotech Startup Accelerator) and E14 (MIT Media Lab Startup Accelerator). Julie is a proud fellow of the *Coaching Fellowship program for Extraordinary Young Women Leaders of Impact* and Biotechnology Faculty at Singularity University Canada. Julie is also the illustrator for this book.

Junior Editors

@PatriciaRea20

@pau_anta05

Patricia was in Grade 7 when she began her role as Junior Editor on *Zero to Genetic Engineering Hero* (2018). She had just finished competing in her first regional science fair, where she won a Bronze Medal for using CRISPR to engineer yeast with a green fluorescent protein and testing it for survival at different temperatures. Yet, her curiosity about genetic engineering began years before, at age 7, after seeing *How To Train Your Dragon* and hearing about geneticists looking to bring back the woolly mammoth. This sparked questions like "What would it take to create a dragon genome?"

Zero to Genetic Engineering Hero gave Patricia the foundation to go deeper into synthetic biology essential lab and biosafety skills, and to gain the understanding required to begin reviewing and analyzing research papers. For her Grade 8 science fair project, she engineered genes to express antifreeze proteins (from the Eel pout and Spruce budworm) into yeast, and tested them for survival at temperatures as low as -196 $^{\circ}$C. This won her First Prize at the 2019 *Lunenfeld Tannenbaum Research Institute Science Fair*, a Gold Medal at the *York Region Science & Technology Fair*, and a Silver Medal at the 2019 *Canada Wide Science Fair*, where she also received the *Discovery Award* for Canada's top research project by an elementary student.

In 2019, Patricia was a panelist at MIT's *Biosummit 3.0*, where she also presented her research on antifreeze proteins. She has twice been selected to Team Canada at international science fairs: *Expo Sciences International 2019* in Abu Dhabi, UAE, and *MAGMA Exporecerca Jove International Research Fair* 2020 in Barcelona, Spain.

Patricia volunteers with Ontario's youth-led group STEM Kids Rock, and is working towards setting up the first high school iGEM team in Ontario. Her long-term goal is to create a strain of yeast that can achieve cryptobiosis to survive the temperature, atmosphere, and radiation of Mars. She lives in Markham, Ontario, with her parents and younger brother, and many Petri dishes of yeast.

Pau Anta (2005) is a maker at the intersection of digital and life sciences, who believes that genetic engineering and synthetic biology can solve some of the grand challenges of our time. He engineered cells for the first time at age 12, using Amino's technology and training, and is interested in the development of practical bio-based applications.

For years now, Pau has been doing a STEM project every year. His most recent one being the development of a deep-sea dropcam, a low-cost research device to capture video footage of biodiversity on the dark seafloor. He presented this project at the *Biosummit 3.0* conference hosted at the MIT Media Lab in 2019.

In 2018, Pau was featured in a documentary about biohacking, produced by Japan's national broadcasting company NHK.

Pau is currently using the skills and knowledge he gained through his *Zero to Genetic Engineering Hero* journey to work on a novel biomanufacturing project involving spider silk, protein engineering and 3D printing.

Pau is a 9th grader at the Washington International School in Washington DC. Besides STEM, his interests are manga, The Matrix, and water polo.

Preface

The world of biology is on the cusp of a tectonic shift. At present, biology is widely thought of as a topic of discovery, of something that needs to be further understood. Biology is also often thought of as an immovable force that marches forth supreme. But over the last two decades, a new mindset has emerged where biology is instead seen as a powerful platform technology for "making". Biology-as-a-Technology (BAAT) mindset, is an engineering mindset where biology is seen as a problem-solving tool that will enable us to solve big problems such as health, ageing, hunger, energy, and even communications.

The change toward the BAAT mindset is profound. However, the tectonic shift is beginning because of the introduction of technologies such as Minilabs, which enable non-experts to start learning and innovating within the BAAT mindset.

Learning and doing genetic engineering has traditionally been exclusive to experts in universities, colleges, technical institutes, corporations, or government. However, a new age has begun where this exciting BAAT mindset, which can result in breakthroughs that save millions of lives every year, is now within the grasp of a much wider audience. Enthusiasts, students, and non-experts can now learn and practise genetic engineering in their home, school, or after-school club and begin their journey to impact the world ten years earlier - this has profound implications on the health of the planet and our civilization.

This book explores the world of biology through the BAAT mindset. Reading this book not only results in learning how biology functions but also how the functions of a cell can be harnessed to make fun or important things. Throughout this book, you'll learn the ins and outs of growing and genetically engineering bacteria, the foundational skills that professional genetic engineers use daily. Upon mastering the content and exercises in this book, you will be a genetic engineer. You will have the first principles, skills, and confidence to explore genetic engineering further into more sophisticated realms.

The learning approach in this book begins with simple exercises. Through the many chapters and many more hands-on activities, you will start to see genetic engineering as trivial. This is when you know you're becoming a Genetic Engineering Hero

The genetic engineering exercises in this book are fantastic tools to learn principles and demonstrate your capabilities. The Nobel Laureate Richard Feynman stated this wonderfully when he said - **"What I cannot create, I do not understand."**

In the world of genetic engineering, the synthetic biology pioneer and Stanford University Professor Drew Endy once said: **"Creating something is the shortest path to demonstrating what you know and what you don't."** You will be engineering cells to make many things, and we are confident that this will help demonstrate your new-found knowledge of genetic engineering.

We thank you for embarking on this journey to understand engineering biology more deeply. We hope that wherever this learning takes you - pioneering new life-saving therapies, making more ecologically-friendly manufacturing processes, making sound investments, developing a new kind of art, or merely helping society to better understand genetic engineering, GMOs and becoming "bio-literate" - that it broadens your horizons and is a tool in your life's toolbox that drives success.

If there is any doubt that engineering biology has not altered the course of humanity for the better, we need to look no further than to XKCD:

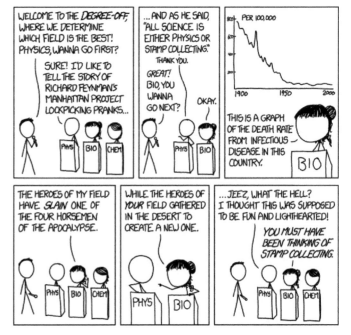

© XKCD. Permanent link to this comic: https://xkcd.com/1520/

Who this book is for:

This book is designed and written for hands-on learners who have little knowledge of biology or genetic engineering. This book focuses on the reader mastering the necessary skills of genetic engineering while learning about cells and how they function. The goal of this book is to take you from no prior biology and genetic engineering knowledge toward a basic understanding of how a cell functions, and how they are engineered, all while building the skills needed to do so.

By the end of this book, if you have completed and practiced all the exercises, you will have university-level lab skills in the area of genetic engineering.

This book is not a typical textbook!

While you will find depth to the content in this book like you would in an average textbook, *Zero to Genetic Engineering Hero* is made to provide you with a first glimpse of the inner-workings of a cell. It further focuses on skill-building for genetic engineering and the Biology-as-a-Technology mindset (BAAT).

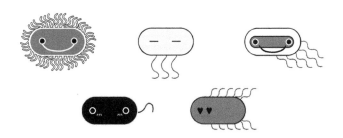

Acknowledgements

This book has been in the works for several years. It is the result of a significant change in the way we think about biology, the fundamental tools that are needed, and how to offer a much improved and more engaging method to teach and learn about biology.

To make the Zero to Genetic Engineering Hero book possible, a new technology ecosystem needed to be invented, refined, and deployed. Novel user

experience methodology needed to be developed - for this, we thank the Amino Labs team and advisers. A special thank you to Trinity Deak, Josh Friesen, Grace Young Kwon, and Safwan Akram for their help in finalizing this edition of the book.

While this book is a departure from what it was initially thought to become, we want to thank Connor Dickie and Dr. Ellen Jorgensen for contributions in an early project with O'Reilly Publishing that did not come to fruition. This was none-the-less an important part of the journey.

We would like to thank and acknowledge many who have helped change the mindset of what genetic engineering and synthetic biology are. In particular, members of the Synbiota Inc. team, including Connor Dickie, Mason Edwards, Pantea Razzaghi, Alex Meunier, Cliff Lau, Heather McGaw, Vipal Jain, James Phillips, Britt Wray, and Dave Russell who worked feverishly in the pursuit of understanding scientific accessibility to close the gap between scientists and the general population. In the same vein, several institutions can be considered pioneering supporters of the biorevolution - Real Ventures, Joi Ito, Exponential Biotechnology LLC, Haliburton County Development Corporation, E14 Fund, the Shuttleworth Foundation, GrowthX, Pinetree Capital, Sheldon Inwentash, and SOSV.

Early support for the development of the Amino Labs technology ecosystem was invaluable. Thank you to Dr. Natalie Kuldell, Dr. David S. Kong, Alexis Hope, Dr. Raymond McCauley, Will Bunker, Jorge Conde, David Strand, Habib Haddad, Calvin Chin, Brett Proud, Stacie Slotnick, Janine Liberty, J. Philipp Schmidt, Kevin Slavin and the Playful Systems group, Andrew Lippman, Lorena Altamirano, Ryan Bethencourt, Arvind Gupta, Michael Hirsch, Stefanie Friedhoff, Jamie McIntyre, David Hu, Kamal Farah, Felix Heilbeck, Basheer Tome, Cory Schmitz, Deepak Jagdish, Chris Peterson, Bianca Datta, Savannah Niles, Sands Fish, and Robocut Studio.

We thank Rafael Anta and Jim Rea for being both fantastic parents in supporting their children's education, as well as for all the feedback and help with refining content in this book.

We are very thankful to our families, friends, and colleagues for their endless support.

Genetic Engineering Hero Community

Visit www.amino.bio/community
to ask questions, share your work and experiment results,
see breakout exercise, review question answers
and meet other Genetic Engineering Heroes!

The Genetic Engineers' Pledge

For the betterment of humanity, I pledge,
With all my DNA, cells, and knowledge,
To never use my genetic engineering mastery,
To lay harm on the natural world or anybody.

_____

To take your pledge, sign & date above.

Chapter 1

Isolating DNA, the Blueprints of Life

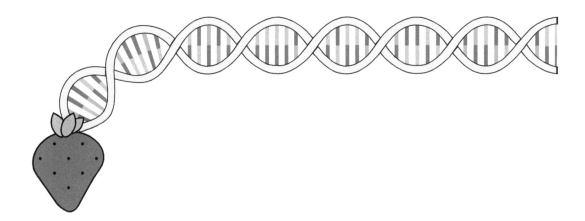

Zero to Genetic Engineering Hero is unique in many ways. So before we jump into the topic of Chapter 1, let's learn how the chapters in this book are structured.

You will notice that this book is packed with hands-on exercises. Each chapter starts with the hands-on activities so that you can "pull up your sleeves" immediately and start building genetic engineering skills. Within the hands-on exercise sections, you will find "Going Deeper" and "Breakout" sections that will provide more information about what is going on in the hands-on exercise. You can read them or skip them - you decide how much you want to learn. At some points, you may feel like you need more background information, but trust us and continue with the exercise to completion as there is a wealth of information later on.

Each chapter has a *Fundamentals* section which covers the first principles at play in the hands-on exercises. These fundamentals are deliberately placed after the hands-on exercises so that you can use real-world skills and context to anchor the concepts in the *Fundamentals*. If, during the hands-on, you would prefer more context, you can always start with the *Fundamentals* and circle back to complete the exercises.

In Chapter 1, you start your Genetic Engineering journey with a simple, engaging hands-on exercise: extracting DNA (deoxyribonucleic acid) from fruit. You've likely heard of DNA before in classic movies like Jurassic Park, forensic television shows, or in the news. But very few individuals have ever seen DNA in real life or know how to extract it from a living or dead organism. As you learn how to extract (separate) DNA from their cells and see it with the naked eye, you will also be learning the general principles of breaking cells open & extracting DNA. You'll likely be surprised at how much DNA you consume when you eat fruit!

This exercise will help you to more richly understand the knowledge gained in the *Fundamentals* section, which focuses primarily on DNA, its structure, and its function. Further, by completing the hands-on exercise, you will begin laying the foundation for understanding and doing genetic engineering in the chapters to follow.

For this chapter, you do not need any special equipment, and you can do the exercise in your kitchen or at your desk. Chapter 2 will cover setting up the necessary equipment and space for the exercises in later chapters.

Getting Started
Equipment and Materials

You do not need any special equipment for this exercise. The materials can be obtained either by purchasing a **DNA Extraction Kit** online at www.amino.bio or by purchasing the components listed below at a local grocery store or pharmacy. The last three items (in bold) are not included in the **DNA Extraction Kit.**

If you are eager to complete all of the exercises in this book, you can purchase all the hands-on kits in packs. These *Zero to Genetic Engineering Hero Kit Packs* are separated into a beginner pack: *(Ch. 1-4)* and advanced pack *(Ch. 5-7)* and are available for both home and the classroom. Find them at www.amino.bio

Shopping List

Water (distilled, or bottled water) - 1 tablespoon
White salt - 1/4 teaspoon
Translucent (non-creamy) shampoo, hand or dish soap with EDTA (ethylenediaminetetraacetic acid) - 1/4 teaspoon
1 paper coffee filter (#4 or #5 are a good size to fit in a cup)
1 very narrow glass, such as a shot glass
1 small sandwich resealable bag
91 to 99% isopropyl alcohol (rubbing alcohol) - 2 tablespoon (70% isopropyl alcohol will also work if you cannot find the higher %)
1 small drinking cup
1 soft fruit (1 strawberry is best but ½ a kiwi or ⅛ banana can work)

Instructional Overview

1. Make a salt water solution in a resealable bag
2. Mash up fruit in salt water to separate the fruit into individual cells
3. Add soap to lyse (cut open) the cells to release the DNA
4. Filter the lysed cell debris to isolate dissolved DNA
5. Precipitate (separate) the DNA in alcohol so it becomes visible

Chapter Timeline Overview

Timeline to complete the hands-on exercise is:

Day 1: ~15 minutes for the Breakout Exercise, ~20 minutes for DNA extraction

Timeline to read *Fundamentals* is typically 3 hours.

Keep in mind!

When doing genetic engineering or any life science project, the activities boil down to chemistry. As you go through this and future exercises, pay attention to the *Going Deeper* sections and to how you are using chemistry to manipulate the biological systems. The more you understand the rules of chemistry, the better you will be able to engineer biology. If there is a scientific term you do not understand, use the internet to learn about it!

What is DNA? *Breakout Exercise*

While you wait for your DNA Extraction Kit or supplies, you can prepare with some contextual exercises on DNA and get going now!

Grab your laptop or go to your desktop computer and visit www.amino.bio/vbioengineer to access the *What is DNA?* virtual simulator. Note that the virtual simulators are not compatible with mobile devices.

What is DNA? is a simple exercise that will provide you with basic context and historical facts about DNA in relation to the genome. You can do the exercise as many times as you like or need, and at any point during this book. We've found that a few tries at the drag-and-drop exercise really helps grasp what genomic DNA is.

During this chapter's hands-on activity, you will be getting a piece of fruit like a single strawberry and mashing it up so that the cells that make up the fruit become separated. You will then break open these cells to release the inside components into the outer environment. This includes releasing the cell's DNA! Finally, you will use some simple filtration and chemistry to cause the cells genomic DNA to become visible to the naked eye.

In other words, you will be extracting the genome from the strawberry cells, so knowing what a genome is will clarify what is happening in your experiment!

Find out about genomes now with *What is DNA?* This will also prepare you to grasp the more in-depth information found in the Fundamentals section.

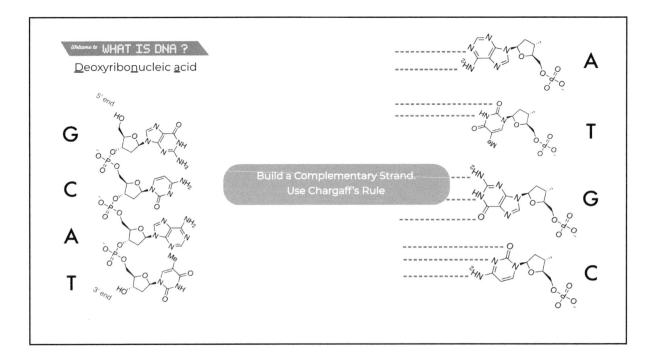

Learning Hands-On:
Breaking Cells Open & Extracting DNA

Step 1. Create salt water inside the resealable bag

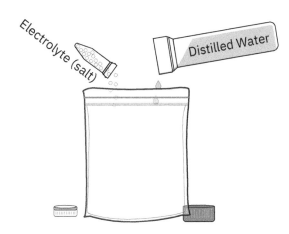

Figure 1-1. Step 1. Create salt water using distilled or bottled water.

In a small resealable bag, mix together **1 tablespoon of distilled or bottled water** with **¼ teaspoon of white salt.** The distilled/bottled water will help you in creating a liquid slurry of mashed up fruit, ultimately enabling you to create a suspension of separated fruit cells. It is important to use distilled or bottled water because tap water contains lots of salts and other impurities that could ruin the experiment.

Why these ingredients? *Going Deeper* **1-1**

Why distilled or bottled water? Lots of calcite ($CaCO_3$) and dolomite ($CaMg(CO_3)_2$) are often found in tap water and as you'll see in coming chapters, calcium (Ca^{2+}) and magnesium (Mg^{2+}) ions play an essential role in how DNA is regulated, driving molecular interactions, and even controlling many biochemical reactions that happen in cells. Moreover, there are often chlorine and fluoride in water for preventing the growth of bacteria and other organisms. While your DNA extraction may still work using tap water, it will work better with distilled or bottled water which has reduced ions (ions are charged atoms like Ca^{2+}).

Why add salt? Salt is added to help the DNA stay separate from the cellular machinery and macromolecules that make up the fruit slurry. When you break open the cells in the next steps, there will be thousands of different molecules and ions floating together that will want to bind to one another in a jumble. By adding table salt, which is mostly sodium chloride (NaCl), the salt binds to and create "buffers" or shields around many of the molecules, including the DNA. Just like the bumper on a bumper car that causes you to bounce off another car during a collision salt ions become the bumpers around DNA. In other words, the salt helps keep the DNA free from binding with (sticking to) other molecules. This enables you to get a larger quantity of 'pure' DNA at the end of the exercise.

Step 2. Mashing the strawberry into individual cells

Figure 1-2 Step 2. Mash the strawberry to separate the cells that make up the strawberry.

Figure 1-3 Visualizing Step 2 of DNA extraction protocol.

You will now add **one strawberry** (or ½ a kiwi, ⅕ of a banana) to the salt-water solution. Mash up the fruit by massaging it in the resealable bag until it is a smooth fruit slurry with no pieces or lumps. (You can keep or remove the leaves, it is your choice.) At the end of this step, the fruit cells will have been separated from each other (Figure 1-2).

These individual cells are now "suspended" in the slurry. Most of the cells will still be intact and functioning, but some will have been torn open by the mixing.

The strawberry is made of millions of individual cells that are tightly packed together, each containing genomic DNA. By separating the cells as much as you can, it will be easier for the soapy chemicals used in Step 3 to come into contact with each cell and cut them open (Figure 1-3).

Extracting DNA from other organisms *Going Deeper* **1-2**

If you want to extract DNA from other organisms, similar principles are applied. You collect a sample such as a leaf, some other fruit, suspend it in a salt solution, break the cells apart, and continue the following steps in this exercise. You can even collect your cells from your mouth by gently scraping them with a utensil and depositing them into a cup or small resealable bag and follow the same procedure.

Some cells have solid outer membranes/cell walls (like yeast) and could require further chemistry or heat to be broken open. You are likely familiar with the COVID-19 virus; in this case, the genome is made of RNA (Ch. 4). Similarly, this virus can be broken open using surfactants. The RNA is then accessible for testing!

Step 3. Breaking open the cells with soap

Figure 1-4 Step 3. Breaking open the cells using surfactants and EDTA.

Now we need to break the cells open, a process scientists called lysing. This process will release the DNA from being contained in the cell.

Add **1/4 teaspoon of shampoo/soap with EDTA,** also called **Lysis Buffer** into the fruit slurry and mix for 3 minutes. If Step 2 was not completed successfully and cells were left in clumps, lots of cells inside the clumps would be protected from coming into contact with the cutting power of the shampoo, resulting in less freed DNA. When the DNA has been released into the saltwater environment, it remains dissolved.

EDTA & Surfactants *Going Deeper* **1-3**

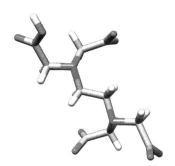
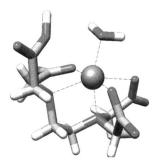

Figure 1-5. EDTA molecule before (left) and after (right) it binds up a metal ion (orange sphere). Dashed lines indicate the bonding between the metal ion and EDTA. By binding most metal ions, proteins no longer bind to DNA. The small red and white V-shaped molecule on the right that is also bound to the metal ion is a water molecule.

EDTA or Ethylenediaminetetraacetic acid (also called Ethylenediaminetetraacetate) is a small molecule made of carbon, nitrogen, oxygen, and hydrogen that is really good at binding to positively charged metal ions like calcium (Ca^{2+}), magnesium (Mg^{2+}), zinc (Zn^{2+}) and even iron ($Fe^{2+/3+}$). It does this by sandwiching the metal ion between its four "arms" very tightly. Have a look at the periodic table at the end of the book and try to find these chemical elements!

During regular cell operation, DNA is bound to cellular molecules called proteins, with the help of metal ions. The proteins are essential for cell operation as they read and copy the DNA. Our goal in this experiment is to get pure DNA. Therefore, we have to remove the proteins bound to the DNA.

The proteins cannot easily bind to the DNA on their own - they have help from positively charged metal ions like magnesium (Mg^{2+}) and calcium (Ca^{2+}). The Mg^{2+} and Ca^{2+} act as a glue which binds to the proteins and the negatively charged DNA (Figure 1-6). If we can remove the metal ions, then we remove the proteins! EDTA is a great way to do this because it binds so tightly to the metal ions. This is also why EDTA is added to shampoo and soaps, it helps to clean you by removing metal ions that bind other molecules to your body. See Figure 4-14 in Chapter 4 to see a similar phenomenon when Ca^{2+} causes DNA to interact with the surface of cells during genetic engineering.

In addition to EDTA, shampoos and hand soaps have surfactants inside. Surfactants are the molecules that cause bubbles to form. A common surfactant is called sodium laureth sulfate (SLS). Surfactant molecules can bind to many different molecules and sequester them. They have a hydrophobic end ("water-avoiding") that can bind to hydrophobic molecules such as fats and oils, and a hydrophilic ("water-loving") end that can bind to other hydrophilic molecules such as water, proteins and

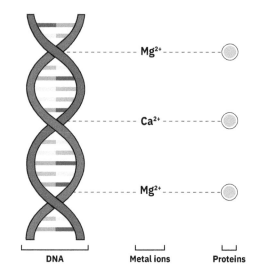

Figure 1-6. Metal ions are like a 'glue' holding the (-) charge DNA and the proteins together.

sugars. A general rule to consider when thinking about how molecules interact is "like binds to like". In other words, molecules that are water-avoiding will generally 'like' to interact with other water-avoiding molecules.

When a surfactant is added to the cells, the surfactant "attacks" and cuts into the membrane of the cells (Figure 1-7; Step 1). As the cell breaks into pieces, different cell parts will interact with the surfactants as they form micelles and get parceled away in the micelles (Figure 1-7; Step 2). The same thing happens when you wash your hair! Dirt, grease, and metal ions are gobbled up into the micelles, but your hair is strong enough to hold up against the cutting power of the surfactant.

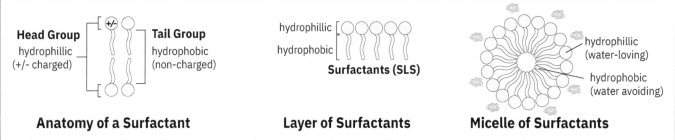

Figure 1-6. The surfactant "SLS" has a charged head that can interact with watery environments and a non-charged tail which doesn't like to interact with watery environments and other charged heads but like to interact with other non-changed molecules. This results in the surfactant molecule heads interacting with one another and the normal watery cell environment, and the tails interact with each other. Ultimately they form a micelle, a spherical ball with the inside being the surfactant tails and the surface being the charged head group.

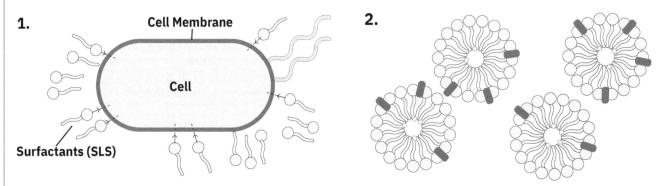

Figure 1-7. (1) The soap molecules called surfactants, attack and cut into the cell membranes of the fruit cells; (2) The surfactants then form spherical jumbles called micelles which contain surfactants and cell debris.

Step 4. Filtering the cell debris

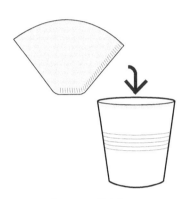 

Figure 1-8. Step 4. Filtering cell debris.

You now want to separate the DNA that is still dissolved in the salt water from the rest of the cell "debris" and micelles that you are not interested in for this exercise. This debris includes the carbohydrates and lipids from the membrane of the cells, soapy micelles with cell debris, as well as clumps of proteins. Simple filtration will be sufficient to separate this debris and micelles from the dissolved DNA.

Set a **paper coffee filter** in a cup and pour your bag's contents into the filter (Figure 1-8). Allow a few minutes for the salt water liquid containing DNA to pass through the filter. As the water, salt, and dissolved DNA pass through the filter, accompanied by some excess shampoo components, a small amount of proteins, and some colour pigments, the fruit's remaining molecules will get caught in the filter. At the end of this step, you'll have dissolved DNA in slightly soapy salt water in a cup. You need only ~½ teaspoon of filtered "DNA" for the next step as there is plenty of DNA in that volume of liquid. Note that if you add too much-filtered DNA liquid in the next step, the precipitation may not work as well - the next section will explain why.

You will notice that the DNA liquid looks like red water. How do you know that DNA is actually in there? Let's make it visible using a chemistry technique called a precipitation!

Step 5. Precipitating the DNA

In this last step, you are going to use chemistry to cause the DNA to "fall" out of the solution so that it can be seen with the naked eye. This "precipitation" is one of the most commonly used techniques in chemistry. Get your kit's empty tube and fill it with 91% to 99% isopropyl alcohol or add 2 tablespoons 91% to 99% **isopropyl alcohol** to a narrow glass and set it in front of you. You are about to see some chemical wizardry!

If you look at the liquid that passed through the filter, which scientists call the filtrate, you will notice it is clear and tinted red thanks to some red pigment from the strawberry that remained in the salt water. You cannot see any DNA, can you? To make the DNA visible, pour a small amount of the filtrate (~1/2 teaspoon) containing the DNA into the isopropyl alcohol tube or glass. As you pour, you will begin to see the DNA precipitate out of solution. After a minute you'll see a white stringy glob of DNA. It may even end up floating to the surface!

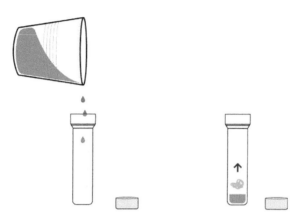

Figure 1-9. Step 5. Pour a small amount of filtered DNA into the isopropyl alcohol.

Molecular interactions *Going Deeper* **1-4**

Why was the DNA dissolved in the salt water? What causes the DNA to precipitate? If you tried the *What is DNA?* simulator, you would have seen that DNA is negatively charged due to the phosphates in the sugar-phosphate backbone. If you have forgotten, or haven't done it yet, have another look at *What is DNA?* Pay special attention to the phosphate molecules - notice how many negative charges are on each phosphate.

In chemistry, "bonding" is a phenomenon where atoms and molecules have a tendency to attract or repel one another. A deeper overview of bonding can be found in Chapter 6. For this exercise, however, you can look to Table 1-1, which describes some simplified rules for whether molecules interact or not. You will see a general theme whereby "like interacts with like". Charged molecules with a positive or negative charge do interact, uncharged molecules do interact, while charged and uncharged molecules do not interact.

Table 1-1 - Simplified rules for molecular interactions		
Molecule 1 Characteristic	**Molecule 2 Characteristic**	**Interaction?**
Charged (+ or -)	Charged (+ or -)	Likely
Uncharged	Charged (+ or -)	Unlikely
Uncharged	Uncharged	Likely

Because both DNA and water are charged, they interact with one other. Before pouring the DNA into the alcohol, the DNA is fully surrounded by, and is interacting with water molecules. The water molecules are able to bond to and "hold onto" the DNA. When the DNA interacts with the water, the DNA is said to be dissolved. You'll learn about bonding later on in Chapter 6.

When you poured the dissolved DNA into the isopropyl alcohol, something different happens. Isopropyl alcohol is an uncharged molecule, and so it does not like to interact with the DNA. As you poured the water-DNA into the isopropyl alcohol, the water molecule shell surrounding the DNA can move away into the isopropyl alcohol. Because water is partially charged, unlike DNA which is very charged, water can mix and partially interact with the alcohol. This means the water does not need to stay bonded to the DNA.

Further, a process called diffusion causes water to move away from the DNA throughout the tube of alcohol. Because there are fewer and fewer water molecules dissolving the DNA, and the DNA molecule does not like to interact with the uncharged isopropyl alcohol molecules, the DNA begins to fold upon its charged self with help from the (+) charged salt you added, also known as sodium (Na+). The folding DNA becomes larger and more dense to the point where it becomes visible to the naked eye and "falls out of solution". When this happens, the DNA is said to precipitate.

Imagine you are at a massive concert. You are standing on the floor and are surrounded by thousands of people who are clapping, singing, and cheering. The singer dives off of the stage into the crowd and begins crowd surfing on the electrified audience. Imagine that the singer is the DNA and that the crowd, you included, is a sea of water molecules. The arms of the water people in the audience are the bonds that can hang onto the DNA singer. As the DNA singer surfs across the crowd, you put your arms up into the air, and the singer passes over top of you. Your arm bonds can connect to the singer and temporarily hold onto the singer keeping them in the air. The DNA molecule singer coasts across the arena because all of the water molecule bonds are keeping the singer held up "dissolved". This is what is happening in the DNA/salt/water solution; the DNA is dissolved.

What would happen to the DNA if the water bonds disappeared? What if all of a sudden the crowd of people was replaced by thousands of cats? Let's pretend that the cats are the isopropyl alcohol molecules. Cats have no desire to hold up people, even a world-famous singer at a concert. Of course, the cats wouldn't be capable of holding up the singer, even if they wanted to. The singer falls to the ground as the cats scatter away to find a quieter more relaxing environment. In this scenario, the singer (DNA) would precipitate onto the ground in the sea of cats (isopropyl alcohol molecules).

What if we slowly added cats to the concert? As cats wandered in, they'd begin filling in space between the people. But as more and more cats pour in, they start causing the people to separate. No one likes to step on a cat, so the people accommodate the cats that are taking over the auditorium and separate from each other more and more. As the people become further apart this makes the crowd surfing singer more nervous because the water audience and the arm bonds are no longer blanketing the auditorium - there are patches of holes forming which the singer might fall through. The DNA singer curls up into a ball to brace for a fall. This makes the situation worse because whereas before the singer was spread out and many arm bonds could hold them up, now they are more compact and dense and fewer arm bonds can touch them and hold them up. With the high number of isopropyl alcohol cats and the low number of water people, the curled up DNA singer falls to the ground - they precipitate out of solution.

This happens because:

- The isopropyl alcohol cats don't have the arm bonds to dissolve the charged DNA singer
- The cats caused the water people to spread apart, lowering the arm bonds that could hold onto the singer
- The DNA singer curled up into a ball and became denser

The cat analogy isn't a perfect one, but it does drive the point across. The non-charged isopropyl alcohol molecules are not keen on interacting with partially charged water or the charged DNA. Because DNA is a charged molecule, and isopropyl alcohol is not charged, only weak interactions occur, and the following occurs as you pour:

In the filtrate, many water molecules surround the large DNA molecules and hold them dissolved. When you pour the water/DNA filtrate into the isopropyl alcohol, the water molecules surrounding the DNA become spread out in the isopropyl alcohol leaving the DNA "naked". The isopropyl alcohol and the DNA do not like to interact with one another so the DNA folds up and binds to other parts of itself with the help of sodium (white salt) and becomes denser.

The large DNA molecule will continue to fold upon itself into large compact clumps of DNA. When the clumps become large enough, the DNA becomes visible, and it also can no longer be dissolved in the isopropyl alcohol. It begins to precipitate out of solution.

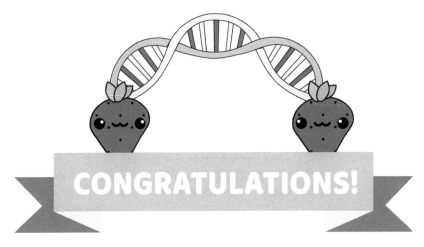

CONGRATULATIONS!

on completing the first experiment!

Were you surprised at how much DNA you found in a single strawberry? You added only a small amount of DNA to the alcohol, and yet there was a lot of DNA. Now consider all of the other food you eat - if the food was at some point living, then it will have DNA in it - you eat a lot of DNA on any given day!

Now that this experiment is complete, you won't be using your isolated DNA further. You can keep it or throw it away. Since there are no dangerous or living materials involved, you can flush the liquid part down the toilet/sink and recycle or discard the tubes according to your local rules.

Remember that this experiment was significant for several reasons:

- It helped you begin your journey learning about cells and molecule bonding

- It helped you learn about how to break open cells, which will be used later in this book to break open cells that you have genetically engineered

- You saw DNA for the first time and, as you read the following *Fundamentals* section, you should keep remembering the DNA you saw.

Note that genetic engineers do use this technique to isolate DNA from the environment to use in their projects. You'll learn more about this as you become a Genetic Engineering Hero!

In the following sections, you will take a more in-depth look at the fundamentals of DNA, and how DNA relates to cell survival. You'll also be exploring how DNA can change naturally or through human intervention. This will set the stage for later chapters in which you will start manipulating the genome of cells. Once you have gone through the rest of this chapter, we recommend that you go back and repeat the hands-on learning exercise once more - not only will it reinforce what you've learned, but you will start seeing strawberries, soap, and salt in a way you never have before.

Fundamentals: DNA

Evolution: It's natural for DNA to change

All living organisms have DNA. DNA is like the blueprint or the instructions for a living cell. As you'll see throughout this book, in addition to DNA, living cells are packed full of "nano-machinery" that can read the DNA and "execute its instructions." The instructions embedded in DNA are critical for the cell's survival.

Cells and DNA can be compared to a computer and its hard drive. The computer hard drive holds all the information that is necessary to "run your computer." The information is stored as tiny magnetic dots and dashes in the hard disk. Like Morse code, the little magnetic dots and dashes are the "language" that the rest of the computer can understand. Perhaps the most essential information on the hard drive is the operating system, like Windows or Mac OS. This information includes all images and sound files that make up what you see and hear when your computer boots up, as well as the software that makes your keyboard, mouse, screen, webcam, and microphone work. Without a hard drive, and the dot and dash information it contains, your computer would not function.

The computer has many parts other than a hard drive. These include your screen, keyboard, mouse, cooling system, graphics, central processing units, battery, and more. All of these "parts" together are required to "bring your computer to life" when you press the power button. The hard drive needs the other computer parts so that the information stored on it can be accessed, read, and executed. The computer parts also need information on the hard drive so they can operate and communicate with each other. By changing the dot and dash information in the hard drive, a computer can behave very differently. Over the last 40 years, the operating systems of computers have evolved substantially because they have been repeatedly updated and modified by computer developers.

DNA is simply a different medium for storing information. Rather than tiny magnetic dots and dashes on the hard disk, DNA is a microscopic string of tiny molecules called nucleotides, and it is the order of the nucleotides that make up the language that the cell knows how to read. Also, just as the tiny magnetic dots and dashes on a hard disk can be modified and changed, so too can the string of molecules that make up DNA. In other words, DNA can be altered.

In the natural world, DNA can spontaneously change due to environmental factors, or due to cell malfunction. Environmental factors like ultraviolet light, gamma radiation, chemicals in the environment, viruses, and even molecules produced by other living organisms can all have an "editing" affect on DNA. Just like deleting an important file on your computer can change how your computer operates (or whether it will even turn on), a small change such as deleting some DNA nucleotides can have a profound impact on a living organism.

It is extremely important to realize that DNA is not a permanent chemical. While it is quite stable, DNA is highly susceptible to modification.

The changeability of DNA is the basis of evolution. It is why every organism is unique and why there is such a fantastic array of living organisms in the world around you. For billions of years, the nucleotide sequence making up the DNA of living cells has changed time after time. DNA nucleotides are constantly being erased, duplicated, inverted, jumbled up, and combined with other organisms' DNA. DNA molecules can even be copy-pasted and cut-pasted through chemical reactions!

The genetic history of humans is a great example: Did you know that 5 to 8% of human DNA is made from viruses that once infected our ancestors? An example of a type of virus that can add its DNA to your genome is the Human Immunodeficiency Virus (HIV), which is a class of virus called retroviruses. HIV infects human immune cells, then copies its nucleic acid into human cells so that these human cells will keep this new DNA and make more of the virus. This shows that human DNA has not been solely human since "the beginning." Rather it is the accumulation of DNA from different organisms in nature as well as natural changes to the DNA sequence over time. When enough DNA is transferred, duplicated, removed, or otherwise modified, this leads to an organism that looks and behaves differently from its parent.

It can take thousands, millions or billions of generations for changes in DNA to accumulate to the point where an organism becomes unique. When minor DNA changes happen even one thousand times, the "Great-one-thousandth Grandparents" can look very different than its "Great-one thousand Grandchild." It may even result in a different species! Let's write this out so you can see what 1000 generations look like:

Great great Grand Parent.

That is a lot of grandparents! Do you know who your great-one thousand grandparents were? Do you think they looked just like your grandparents? In humans, 1000 generations of minor DNA modifications require about 15 thousand years. In microorganisms, 1000 generations of minor DNA modifications can happen in days. Considering that a microorganism's lineage goes back billions of years, that provides a lot of opportunities for generations to change and evolve.

Life on earth started with the simplest of life forms, with simple DNA blueprints. Through billions of years and an unimaginable number of incremental changes

to the DNA blueprint, we now see countless species of microorganisms, plants, fungi, insects, and mammals such as humans.

Modern Synthesis

While the understanding of DNA, genetics, and heredity matured over hundreds of years, an essential milestone occurred in the early 20th century as Darwinian evolution and Mendelian genetics transitioned from theory to fact. The combining of Darwinian evolution (natural selection) and Mendelian genetics (inherited DNA) was termed the Modern Synthesis.

Today, Gregor Mendel is known as the founder of genetics, and genetics, of course, is the foundation of the biotechnology industry! But, in the mid-1800s, when Mendel began his scientific inquiries into how traits are passed on from one organism to the next, virtually nothing scientific was known about it. While traditional breeders of livestock and crops had been slowly modifying organisms over at least thousands or tens of thousands of years, scientifically proving that biology could be changed through domestication had yet to be achieved.

It wasn't until Mendel started his pea plants experiments that crucial information on genetics emerged (Figure 1-10). Mendel found that when he cross-pollinated pea plants with different colored seeds, some of those seed colors were preferentially passed down to the next generation of plants. But, looking further down the line of pea plant generations, he found that some seed colors eventually re-emerged! His experimental procedure, and therefore the first genetic experiments, went something like this:

1. Yellow seed pea plants were carefully bred with other seeded yellow pea plants to ensure he had "pure-breed" yellow pea plants.

2. The same was done with green seed pea plants.

3. Yellow and green seed pure-bred plants were then carefully cross-pollinated. This is quite easy to do. Today, you can use a Q-tip and gently rub a blossom of one plant to gather pollen, then rub it in another plant blossom, and lastly, rub it back in the first plant.

4. The plants grew, matured, and produced seeds

5. The color of the seeds were identified and kept for similar future experiments.

Figure 1-10. Pea plant. Photo by Crepessuzette from Pixabay

The pure-bred plants would always produce the expected colored seeds: yellow produced yellow seeds and green produced green seeds. However, when he did the first cross-pollination of yellow and green plants, all of the first generation offspring seeds were yellow! It looked as though some genetic information for yellow seeds was preferred over green (Figure 1-11)!

But, things ended up getting a bit more complicated. What do you think happened when these first-generation yellow seeds were grown into plants and cross-bred? You might expect this preferential yellow genetic information to cause only yellow seeds. But the result included the reemergence of green seeds! And the most interesting part was they appear in a ratio of one green seed for every three yellow seeds. This was the first time that rigorous scientifically derived information about how traits were passed down from a parent to its offspring was available. It was also the first instance that showed some "genetic rules" for which information is passed to the offspring. To explain this, Mendel coined the terms "recessive" and "dominant" genes. The green seeds were the ones with the recessive trait and the yellow seeds the dominant trait.

Mendel's studies broadened beyond seed color into different plant traits, including height, blossom colors, pea pod shapes, colors, and more. In these studies, he found other "genetic rules" existed! Today, our hair color, eye color, skin color, and a lot more are well-known examples of how genetics are passed down from parents to offspring.

Modern synthesis combined Darwinian evolution theory and the scientifically rigorous Mendelian genetics to create a unified framework to understand how biological organisms can change through time. Mendelian genetics, now just genetics, turned out to be the mechanism or the "how" and "why" different genetic material is passed down from parent to offspring. Darwinian evolution allowed scientists to see that genetics applied to all organisms, not just pea plants. Modern synthesis also included some other key mechanisms in which DNA changes occur in organisms and populations:

Mutations: when changes in the DNA sequence happen due to copy errors. While mutations can be increased due to environmental factors, this is often considered independent of selective environmental pressures (natural selection). In most cases, mutations are harmful to an organism; however, in a minority of cases, a mutation can lead to a benefit that can help an organism survive in their environment. In other words, mutations do occur randomly but can become a part of natural selection.

Random genetic drift: is where allele frequency changes in populations. An allele is a DNA sequence (*e.g.* gene) that is very similar to another allele but has a slightly different sequence. Alleles are created due to mutations. As organisms reproduce, these different versions of DNA sequences will "move" through a population. Similar to the pea experiments above where the "yellow seed" and "green seed" alleles "moved" through the pea population as the plants were bred.

Gene flow: is where allele frequency changes due to immigration or emigration to/from a population. For example, if there were only true-bred green seed pea plants in a region and a person or bird happened to bring a yellow pea seed to that region. When that pea plant grows, it's yellow seed allele will become dominant in the region in the coming years as cross-breeding occurs.

A key basis for natural selection, random genetic drift, and gene flow are mutations. Mutations are random and result in different DNA sequences (alleles). While most mutations negatively impact an organism, some can provide a benefit. The beneficial mutations can help an organism survive in a particular environment (natural selection) or can help the organism survive independent of the environment. The alleles can then "move" around and change a population through gene flow and genetic drift. Bit by bit, generation after generation, mutations reshape us, and the world around us as species change or new species emerge.

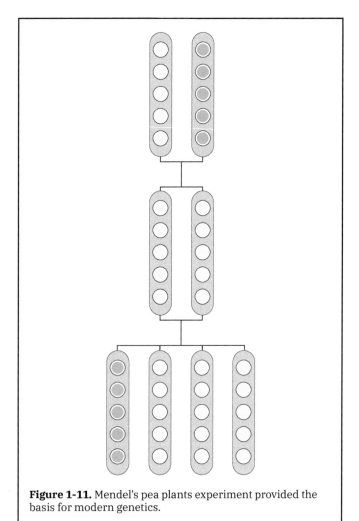

Figure 1-11. Mendel's pea plants experiment provided the basis for modern genetics.

Genetic engineering: The road to precise editing of DNA

Humans have directly influenced the modification of DNA in organisms for thousands of years through domestication and selective breeding. You can see the outcomes of these actions in our homes, farms, and factories. These methods of changing the DNA of living organisms take advantage of evolution. When humans cross-breed plants, the DNA within the plants mix: DNA from one plant will combine with another. The result is modified DNA with new traits that may or may not be beneficial to the plant. With human intervention, however, if the new trait is beneficial and desired by the human, it becomes selected for because the plant's survival will be encouraged by humans through watering, and propagating the plant's seeds. Because of this, plants, especially crop plants such as maize (corn) are almost unrecognizable when compared to their ancestors (Figure 1-12).

Domestication and selective breeding do not only apply to crops. Consider pets for a moment. The "original dog" is *Canis lupus*, better known as the wolf. Dogs have been so extensively bred by humans that a huge dog such as a Great Dane—standing at a minimum of 3 feet and weighing almost 250 pounds—can no longer physically mate and reproduce in nature with a small dog such as a Chihuahua—standing at less than 1 foot and weighing less than 10 pounds. The Chihuahua is 25 times smaller than the Great Dane, yet both are descendants from the same source DNA (Figure 1-13). In the natural world, these two organisms can no longer effectively breed. Their DNA can no longer combine and intermix as would naturally happen in a species. Without human intervention, this will eventually lead to the development of two different species. This is evolution in progress.

The methods in which humans evolve plants and pets through deliberate breeding and selection extends into all corners of the natural world - trees, grasses, flowers, livestock animals, and even microorganisms. Much of the food we eat today was evolved through selective breeding by humans. This process often takes a long time because "DNA mixing" through these methods is random, and desirable traits only appear by chance. What if randomness could be taken out of the equation? What if humans could, with extreme precision, edit DNA?

The precise engineering of organisms is the focus of this book. By its end, you will have gone through exercises where you will have precisely changed an organism's DNA to get a desired trait. Over the past 100 years, humans have made ground-breaking discoveries that have led us to understand what DNA is, how DNA behaves, and its importance in biology. These discoveries have led to new technologies that enable us to precisely read, write and edit the DNA of organisms. There has also been a change in the mindset of people. No longer is nature and the natural world seen as being immovable or unchangeable. A robust engineering discipline has emerged where we can now make very important things using biology. We call this the Biology-as-a-Technology mindset.

The knowledge and skills imparted by this book are not to be taken lightly. Understanding how DNA and cells interact will be vital to making informed decisions about using your new skills. Now that you have seen DNA and know that it can be manipulated let's further explore the building blocks of DNA and the cells where they reside.

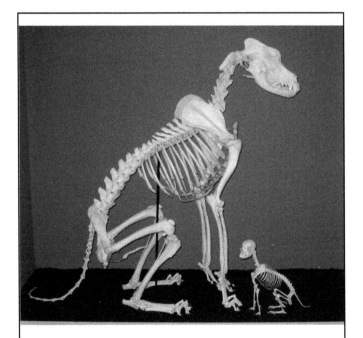

Figure 1-12. Comparison of modern Corn (maize) (bottom), its ancestor Teosinte (top) and a hybrid of both (middle). Photo by John Doebley - http://teosinte.wisc.edu/images.html

Figure 1-13. Great Dane skeleton (left) compared to a Chihuahua skeleton (right), both descendants of *Canis Lupus*, at the Museum of Osteology.

Figure 1-14. A ball pit is a good analogy for thinking about how the inside of cells are packed with molecules.

Atoms, molecules, and macromolecules of the cell

DNA is housed inside cells, but what are cells made of? How does DNA fit in the picture? Let's start with atoms: Are you familiar with the periodic table of chemical elements? You can find one at the end of this book. Take a moment to explore it. The periodic table is color coded to tell you where each of the elements is used in cells. Atoms such as carbon (C), hydrogen (H), oxygen (O), phosphorous (P), nitrogen (N), and sulfur (S) are brought together by machinery in cells to form molecules. A molecule is the combination of two or more atoms that are bound together. Carbon, hydrogen, oxygen, nitrogen, phosphorous and sulfur are generally considered **organi**c elements because they are frequently used by **organi**sms. As you completed the hands-on DNA extraction, keep in mind that most of the "cellular stuff" that you manipulated was made of these elements, and in the case of extracting DNA from cells, DNA is made of CHOPN.

A cell has the necessary machinery to combine atoms into molecules, and it also has the machinery inside to combine molecules together to form macromolecules. The term macromolecule is used to describe the four basic components of cells: **proteins, lipids, carbohydrates, and nucleic acids**. In Chapter 1, the primary focus is on nucleic acids, specifically, deoxyribonucleic acid (DNA) in cells.

Keep in Mind!

Sometimes memory tricks help to remember specific facts. You can try to think of CHOPNS as the word "Choppings." Imagine chopping up some yummy celery or carrots into little bits for your soup. Atoms are like that; the little bits that make up the molecules in cells, just like celery and carrots make up the soup.

Of particular importance in living organisms are the 'organic' elements CHOPNS; carbon, hydrogen, oxygen, phosphorous, nitrogen, and sulfur.

What would it look like if you were to peek into a cell to look at the atoms, molecules, and macromolecules? The inside might look like a ball pit, but rather than colorful plastic spheres, there is a sea of different molecules packed together and bumping into one another (Figure 1-14). Imagine being immersed in the depths of a ball pit - completely surrounded. If you were shrunk down to the size of atoms and molecules - the size of DNA, the colored balls packed against you would be comparable to other molecules of the cell.

The molecules of living cells range in size from very small, such as oxygen gas (O_2) with only two atoms, to water (H_2O) with three atoms, to DNA which is by far the most enormous at millions or billions of atoms. In the DNA extraction exercise, you saw with your own eyes how much DNA was in the strawberry cells - that glob of DNA was made of trillions of atoms. And the other macromolecules, such as the proteins, lipids, and carbohydrates bound up by the surfactants into micelles, were made up of trillions more.

DNA: The blueprints of living cells

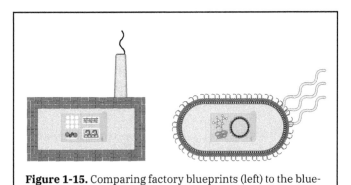

Figure 1-15. Comparing factory blueprints (left) to the blueprints of a living cell, DNA (right).

A blueprint is used for storing information about the construction and maintenance of a factory and the processes inside.

Blueprints for the building may have information about the structural components (*e.g.* brick outer shell supported by iron beams), the ventilation systems (*e.g.* made of aluminum ducting), the complex electrical configurations (*e.g.* different voltages, wires, and location of the outlets). Blueprints can also determine how a building is built. Is a crane needed? If so, scaffolding may be erected to accommodate for this.

Blueprints can also be used to define the equipment inside, how it is positioned, and the speed at which it operates. A lot of equipment and infrastructure is permanently connected to the factory. In this case, the components, machines and the factory building itself meld to become one. In other words, a factory is not just a building, but is the building, equipment, and even the people inside, turning raw materials into end products.

Why are we talking about factories? Factories as we know them are analogous to living cells—except they are much larger and made of different materials. The

blueprints that define how to build and maintain a factory are comparable to the blueprints of living cells. In the case of living cells, however, there is not a miniature piece of blue paper rolled up in every cell. There is a microscopic chemical string, the DNA.

Unlike a factory blueprint that you can see with your eyes, hold in your hands, and is written in a language that is well understood, the blueprint of a living cell, DNA, is a microscopic "chemical string" of nucleotide building blocks. This string has a distinct "language" that all living cells know how to read and write. Only since 1953 when Francis Crick, James Watson, Rosalind Franklin, and Maurice Wilkins discovered the structure of DNA have we begun to understand it and its language. Here is some of what we know:

DNA is made of four chemical building blocks called nucleotides that are attached to one another in various orders to form a "string" of the nucleotides that can be thousands to millions of nucleotides long. Each nucleotide is made of CHOPN, and each nucleotide is made up of three "sub-molecules" called a phosphate, a nitrogenous base, and a deoxyribose sugar (Figure 1-16). Have a look at Figure 1-16, do you see the CHOPN chemical elements?

Phosphate is a molecule with a phosphorus atom bound to four oxygen atoms. Phosphate is a critical component of the "sugar-phosphate" backbone of DNA (Figure 1-17). The phosphate group is very negatively charged and is what gives DNA an overall negative charge. As you saw in Chapter 1's exercise, the negative charge of DNA is an important chemical characteristic that you took advantage of to extract DNA from cells using chemistry.

Nitrogenous bases are the variable part of a nucleotide. While every nucleotide has the same phosphate backbone, there are four different nitrogenous bases in DNA - **G**uanine, **T**hymine, **A**denine, and **C**ytosine (Figure 1-18). Can you spot the different nitrogenous bases? If you have trouble, try the *What is DNA?* application mentioned in the hands-on exercise. The nitrogenous base is the part of the nucleotide molecules that is the information that the cell machinery "reads".

Deoxyribose is a sugar ring that is the "D" in DNA. Deoxyribose connects to both the nitrogenous base and the phosphate (Figure 1-16). When multiple nucleotides are connected, as seen in Figure 1-16, you'll see that the deoxyribose also connects to the phosphate of the next nucleotide. The deoxyribose and the phosphate together form the "sugar-phosphate" backbone of DNA (Figure 1-17).

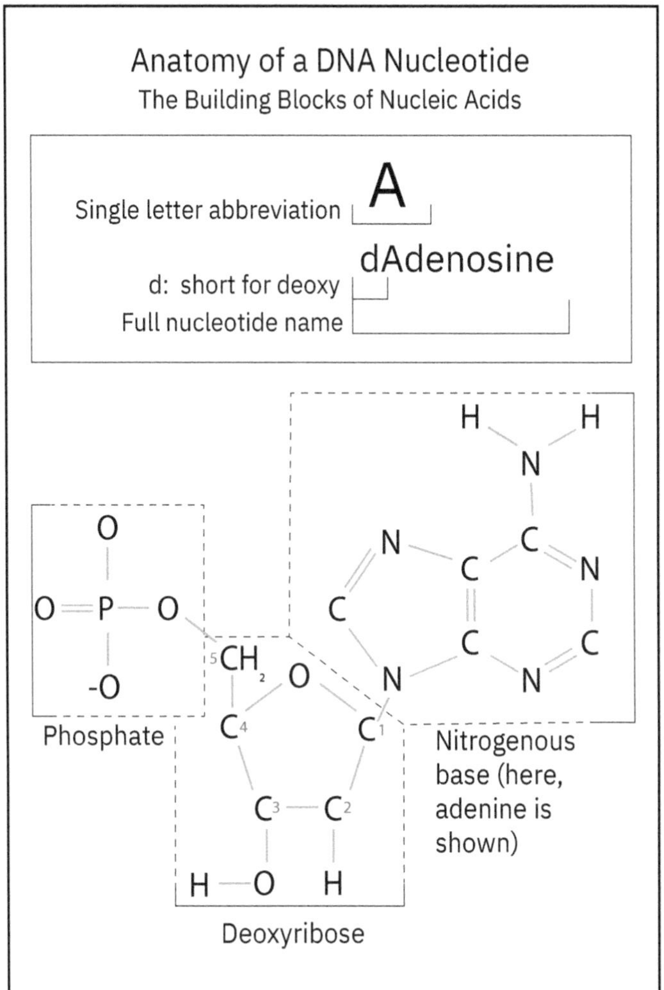

Anatomy of a DNA Nucleotide
The Building Blocks of Nucleic Acids

Single letter abbreviation — **A**

dAdenosine

d: short for deoxy
Full nucleotide name

H   H
 \   /
  N
  |
N   C
 \\ / \
  C   N
  | ‖
  C   C
 / \\ / \
N   N

Phosphate

O
‖
O = P — O
|
-O

$_5CH_2$ O

Nitrogenous base (here, adenine is shown)

C^4 C^1

C^3 — C^2

H — O   H

Deoxyribose

Figure 1-16. The building blocks of DNA, nucleotides, have important characteristics. A conserved phosphate, deoxyribose sugar, and the variable nitrogenous base.

Nucleotides are usually referred to by the first letters of their nitrogenous base names. Because four different nitrogenous bases can be attached to the deoxyribose ring, four different possible nucleotides make up DNA: **A** for **a**denine, **T** for **t**hymine, **C** for **c**ytosine and **G** for **g**uanine (Figure 1-18). You can see two nucleotide chains side-by-side in Figure 1-17, and both have each of the four nucleotides.

During regular cell activity, millions of each of the four nucleotides are produced by the cell. The nucleotides in the cell can be collected by a cell machine called DNA polymerase, which strings them together like pearls in a necklace. It is the order of the nucleotides *...ATGGCGGTTACC...* which we call the DNA sequence that the cell machinery reads and understands.
As you see in Figure 1-17, the sugar-phosphate backbones are on the outside of the molecule, and the nitrogenous bases are buried inside. If you imagine DNA to be a ladder, the sugar-phosphate would be the

outside of the ladder, and the nitrogenous bases would be the ladder rungs. This means that the "information" that the cell's machinery reads is in the rungs of the ladder. If you look at Strand 1 in Figure 1-17, reading from the top "5' Phosphate end" down to the "3'-OH end", the DNA sequence is GCAT.

When two strands of DNA come together, they form a three-dimensional structure known as the double helix. Looking down at the structure from the top, it looks like a spiral staircase (Figure 1-19). The blue-gray region in the center of the illustration shows the nitrogenous bases, the A's, T's, C's and G's that form the information layer of the DNA blueprint. These are held in place by the outer orange-red-gray region of the negatively-charged sugar-phosphate backbone. The sugar-phosphate backbone has two critical functions: i) to maintain structure and hold the nucleotide bases in place; ii) to selectively attract cellular machinery called proteins so that the cell can "read" the DNA. In simplest terms, you can think of the sugar-phosphate backbone as the paper of a blueprint, and the nitrogenous bases as the words (information) written on the blueprint that can be understood.

A cool property of DNA is that the two strands of a double helix are only loosely connected by hydrogen bonds (which you will learn more about in Chapter 6). These loose bonds mean that the two strands can come apart like a zipper on a jacket. On a jacket, each strand of the zipper is quite strong, whereas the interaction between the two zipper strands can be undone easily with a simple pass of the zipping mechanism. In DNA, the chemical bonds in the sugar-phosphate backbone are very strong, but the bonds between two strands are relatively weak (Figure 1-17; dashed lines).

This "unzipping" capability is a key feature of how DNA functions. The cellular machinery can bind to the sugar-phosphate backbone and pull the double helix apart, unzip to get the information as needed, and when finished, the molecule closes up and protects the information.

Are there any rules on how two DNA strands can zip together? Yes! One fundamental rule was discovered by a famous scientist called Erwin Chargaff. Chargaff's rule states that: *Only an "A" in one strand can bind to a "T" in the other, and only a "C" in one strand can bind to a "G" in the other:* the nitrogenous bases of A's and T's complement each other and the same with C's and G's. In Figure 1-17, you'll see Chargaff's Rule in action. Only when two strands have "complementary" nucleotides can they wind up together! As you learn more about the nuts and bolts of genetic engineering in later chapters, this will become much more evident.

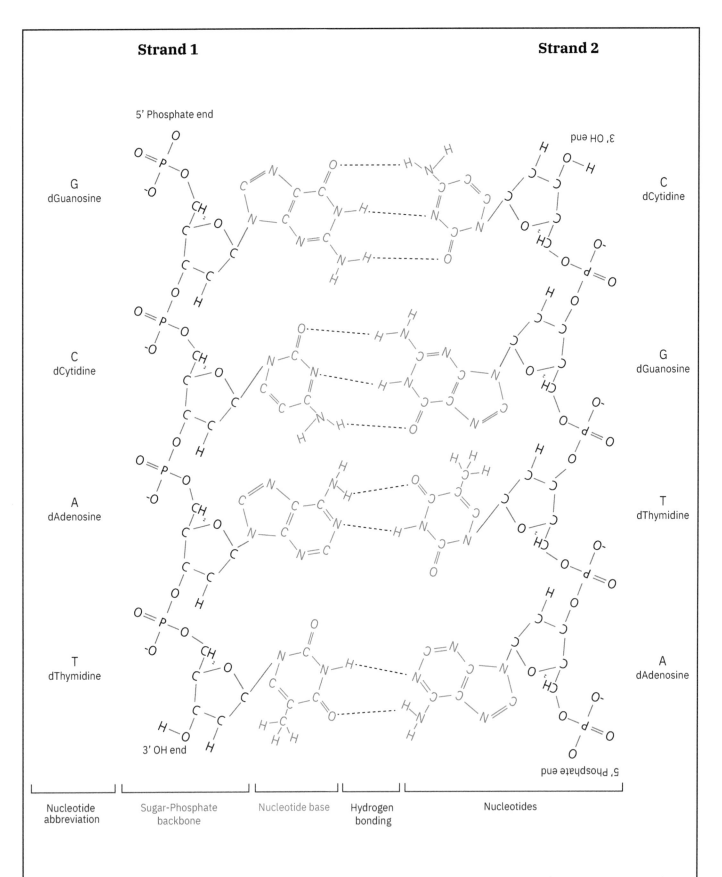

Figure 1-17. Nucleotides are strung together into strands via the sugar-phosphate backbone. Two strands of DNA can zip together to form a double helix. They are able to zip together because of a special bonding force called hydrogen bonding. The hydrogen bonding occurs between the nitrogenous bases alone, not via the sugar-phosphate backbone. You will learn about bonding in Chapter 6.

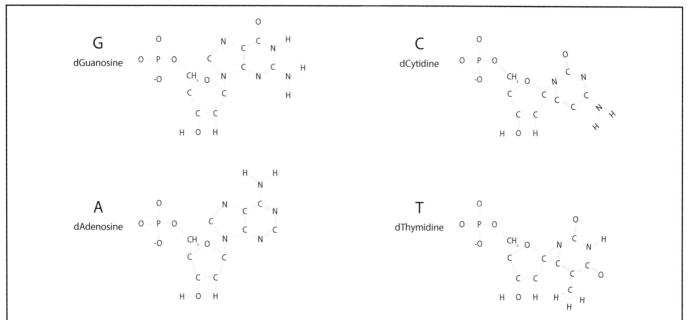

Figure 1-18. The four nucleotide building blocks of DNA are called deoxyguanosine (dG), deoxycytidine (dC), deoxyadenosine (dA), and deoxythymidine (dT). For simplicity, they are often referred to by their nitrogenous base names: guanine (G), cytosine (C), adenine (A), and thymine (T). A small d is an abbreviation for 'deoxy.' You'll learn more about this when you discover RNA in Chapter 4.

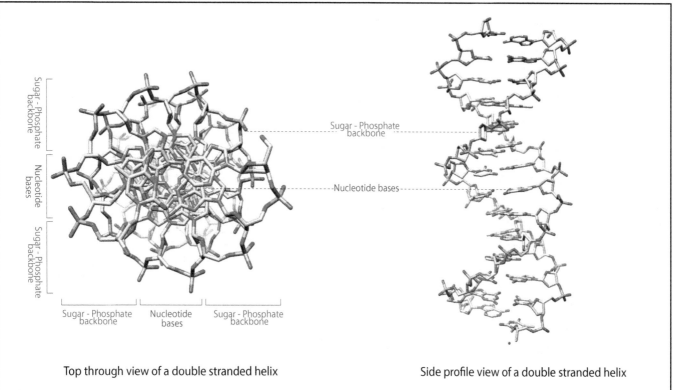

Top through view of a double stranded helix

Side profile view of a double stranded helix

Figure 1-19. The DNA helix is a three-dimensional structure that forms when two strands of DNA bind together using Chargaff's Rule. The blue-gray region indicates the nitrogenous bases, and the red-orange-gray region indicates the negatively charged sugar-phosphate backbone.

Understanding the nomenclature of DNA

In the hands-on exercise of this chapter, you isolated genomic DNA. There are a lot of words that are used to describe DNA. There is DNA, double helix, genes, genomes, chromosomes, genomic DNA and more. All of these can be used to describe DNA at different scales or forms. Let's break this down by looking at DNA from its building blocks up to the "megastructures" that are made from it.

Atoms are the building blocks of molecules, incluuing nucleotides. Earlier you learned that the building blocks of nucleotides are CHOPN. Do you remember what CHOPN stands for? See the periodic table at the end of the book to refresh your memory.

Nucleotides are the building blocks of a DNA strand: The nucleotides adenosine, thymidine, guanosine, and cytidine are the building blocks of a string of DNA. More commonly they are referred to by their nitrogenous base names: adenine, thymine, guanine, and cytosine.

DNA strand: is when several nucleotides become bound together via the sugar-phosphate backbone.

Two DNA strands form a **double helix:** When two complementary strands of DNA join (using Chargaff's Rule), they form double-stranded DNA, and the three-dimensional structure is called a double helix.

A Gene is a segment of double-stranded DNA helix that has all the information required to be read by a cell; which then 'expresses the gene', producing a cellular product or outcome. This expression of a gene can result in a physical change in the organism. A gene can be a 100 to 14,000 nucleotides long. There are an estimated ~25,000 genes in the human genome. This is a major topic of Chapters 4 and 5.

A plasmid is a short circular DNA helix: A double helix that is generally between 1,000 and 100,000 nucleotides long and is circular, is called plasmid (Figure 1-20). Just like a pearl necklace is made of pearl "building blocks" and is circular, the DNA helix can form tiny loops ("necklaces"). Plasmids are a significant topic of discussion throughout this book. They are frequently used in genetic engineering, and cells often share plasmids that help them survive and evolve. In computer terms, a plasmid is similar to a USB stick for cells. It is transferable mobile data storage.

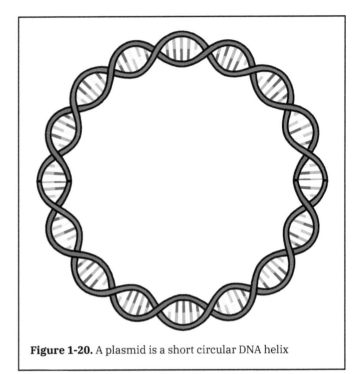

Figure 1-20. A plasmid is a short circular DNA helix

A chromosome is a long circular or straight DNA helix that carries the primary information that the cell uses to survive. Throughout this book, we will be engineering *Escherichia coli* (*E. coli*) bacteria. *E. coli* bacteria have a circular chromosome that is about 4,600,000 nucleotides long. Compare this to a human cell that has 46 straight chromosomes of varying lengths (generally between 10,000,000 and 100,000,000 nucleotides long).

A genome refers not necessarily to a single physical structure in a cell (like a chromosome), but to all of the organism's DNA. Each living cell contains a copy of that organism's genome. In the case of a single-celled bacteria like *E. coli*, its genome is made up of a single large chromosome of 4,600,000 nucleotides and in some cases some small plasmids. In contrast, your human genome includes the 46 (or so) chromosomes that total about 6,000,000,000 nucleotides as well as your mitochondrial DNA (mitochondria is the energy factory within your cells that has its own DNA). When you extracted genomic DNA from the strawberry, this means you attempted to extract all of the cell's DNA.

Negative charge of DNA *Going Deeper* 1-5

Recall during the precipitation exercise that the DNA had a charge. The charge of the DNA makes it "water-loving" because water also has charge. This keeps the DNA dissolved in the water. Have a closer look at the negatively charged sugar-phosphate backbone (Figure 1-17) and the DNA double helix (Figure 1-19). In the DNA double helix, many of the red zones on the outside of the molecule are the negatively charged oxygen atoms of the phosphate group. It is this phosphate group which enabled you to use hydrophilic / hydrophobic chemistry to precipitate the DNA. When you're doing chemistry in general, whether it is biochemistry (the chemistry of biology) or not, you must understand the chemical nature of your molecules to know how to manipulate them. Chapter 6 *Fundamentals* goes in-depth into bonding.

Could your DNA stretch to the moon and back? *Going Deeper* 1-6

If you look back to Figure 1-17, you'll see some DNA strands. What is the distance from the beginning of a single nucleotide (5' P) to the end (3' OH)? It turns out that the length is about 0.33 nanometers or 0.00000000033 meters. Consider humans have about 6 billion nucleotides of DNA in both strands of DNA in each cell. Therefore the DNA in each of your cells is 3 billion nucleotides long. Each of your cells has:

3,000,000,000 (nucleotides long per cell) x 0.00000000033 m = 0.99 m of DNA per cell!

Find out how long a meter is and imagine that each one of your cells, that you need a microscope to see, has 1 meter of DNA neatly wrapped up and packaged inside! If you're wondering how it could be possible to do this, imagine a spool of thread. You can easily hold a spool of thread in the palm of your hand, however, if you are allowed, unroll the entire spool and see how long the string is! Maybe the string will be an entire city block long!

Now for the second part of the calculation. It is estimated that the average human has more than one trillion cells that carry your DNA genome. It is important to note that you probably have around 30 trillion red blood cells, but because red blood cells lose their DNA when they become red blood cells, we cannot consider them in this calculation. For simple calculation, let's say you have one trillion cells with DNA.

0.99 m (DNA length per cell) x 1,000,000,000,000 (cells) = 990,000,000,000 meters!

With this conservative calculation, that is 990,000,000 km of DNA in your body.

The distance between the earth and the moon is about 384,400 km.

990,000,000 km (DNA in your body) / 384,400 km (between earth and moon) = 2,575

Wow! If the DNA in your body was stretched out, it would reach to the moon or back more than 2,500 times!

The distance between the earth and sun is 149,600,000 km:

990,000,000 km (DNA in your body) / 149,600,000 km (between earth and sun) = 6.6 times

Your DNA would stretch between the earth and our sun almost seven times!

The distance to our nearest star neighbor, Alpha Centauri A, is 4.22 light-years away. This is equal to about 13,000 astronomical units (AU). An astronomical unit is the distance between the earth and our sun. How many people's DNA would need to be strung together to reach the nearest star?

13,000 AU / 6.6 = 1,967 people

If you were able to gather up almost 2,000 people, all of their DNA, if stretch out and connected in one long thread, would reach all the way to our nearest celestial star neighbor, Alpha Centauri A!!!

Figure 1-21. Marine divers exploring the oceans for new organisms.

DNA extractions in the real-world

Although tremendous advances have been made in DNA science over the past decades, we are still learning the basics of how biology works. We do not yet have the capability of designing a DNA sequence from scratch to perform a particular function or result in a specific trait. Instead, it is quite common and necessary for genetic engineers, scientists, and enthusiasts to go 'prospecting' in nature for snippets of DNA that make interesting and useful cellular products or cellular machinery.

Just like you extracted the DNA from a strawberry, genetic engineers collect samples of organisms in nature and extract their genomic DNA. They then send the extracted DNA to companies that have a technology called DNA sequencers that can read the DNA to determine the DNA sequence. With the new DNA sequences, scientists can analyze them to find interesting and useful functions which can then be used in genetic engineering projects.

Sometimes this prospecting leads to the discovery of a DNA sequence that results in a multi-billion dollar product that benefits billions of people. For example, Osamu Shimomura, Martin Chalfie and the late Roger Tsien were awarded the 2008 Nobel Prize in Chemistry

for discovering the Green Fluorescent Protein (GFP). The DNA for a fluorescing protein found in jellyfish was collected, analyzed and used in a genetic engineering project. The stories told by the laureates can be found on *nobelprize.org*, notably, *"Nobel Lecture by Osamu Shimomura* (29 minutes)"

It all started with curiosity - what made some species of jellyfish glow a beautiful blue-ish green hue? Osamu Shimomura spent decades working toward understanding this question. In the early part of the research, Dr. Shimomura and his colleagues had to collect jellyfish and sacrifice them to gather both the fluorescing proteins and their DNA. Dr. Shimomura, fellow scientists, and even their family members collected an estimated 50,000 colorful jellyfish from the ocean to analyze the molecules that produce the mysterious blueish green light. Finally, in the 1970s, through much trial and error, they were able to purify enough of the Green Fluorescent Protein (GFP) from their samples to determine how it worked.

The invention of more efficient genetic engineering protocols in the 1990s made it possible to analyze the DNA of the jellyfish and ultimately insert the jellyfish GFP DNA into a bacteria, and have the bacteria produce it. This certainly saved a lot of jellyfish, and it also led to the creation of new and better versions of GFP by changing the original DNA sequence. There are now versions of GFP that are brighter, last longer, and even fluoresce different colors! Fluorescent proteins are now so widely used that thousands of laboratories worldwide depend on them as a research tool to discover how cells work so that medicines may be created. In other words, the discovery of the DNA sequence used to make GFP led to the creation of a multi-billion dollar industry and many new medicines that save lives every day.

In later chapters, you will be able to do this Nobel Prize-winning experiment yourself, putting the gene of a fluorescent protein into a bacteria!

Prospecting for high-value biological products continues today. Medicines from plants and animals, pigments in coral, scents from flowers, and flavors or oils from plants are all examples of products that have been or are being prospected. Prospecting for useful "cellular machinery" is a lucrative business. For example, the biotechnology company *New England Biolabs* (commonly called "NEB"), offers more than 1000 different "molecular scissors" called restriction enzymes to help genetic engineers cut DNA with high precision. Each one of these enzymes was prospected and isolated from naturally-occurring bacteria. Did you hear about CRISPR in the news? CRISPR/Cas-9 was also prospected from bacteria in the environment and has since become a game changer in biotechnology and medicine! The sales from prospected biological products currently amount to billions of dollars in revenue each year.

What can modern-day prospecting look like? Starting in 2003, with his private yacht, genomics pioneer Dr. Craig Venter, began sailing the ocean, looking for industrially useful life forms. In 2004 he and Exxon-Mobil launched the Global Ocean Sampling Expedition, a two-year effort to collect samples from the ocean to "assess genetic diversity in marine microbial communities and to understand their role in nature's fundamental processes". In a more recent 2010 expedition, funding support was provided by Life Technologies Foundation, the nonprofit arm of one of the largest biotechnology companies. We will probably see the fruits of their expeditions in the coming years as critical life-saving technologies that earn their prospectors a handsome salary.

Figure 1-22. Bioluminescent jellyfish. Photo by Smit Patel.

Is DNA the ultimate code? *Going Deeper* **1-7**

You may be familiar with or have seen the binary information that computers use to compute. At the most basic level, computers use switches called transistors which are either on or off - which the computer designates a "1" or a "0". Binary is the language that computers know how to read and write, but it's quite hard for a human to understand!

01100111 01100111 01100011 01100111 01100001 01100001 01100001 01100001 01100011 01100111 01100001 01100001 01100001 01100011 01100011 01100001 01110100 01110100 01110100 01100111 01100011 01100111 01100001 01100001 01100001 01100001 01100011 01100111 01100111 01100011 01100001 01110100 01110100 01100001 01100001 01100011 01100111 01100001 01100001 01100111 01100001 01100001 01100011 01100111 01100011 01100001 01110100 01100011 01100111 01100111 01100011 01100001 01110100 01110100 01100001 01100111 01100011 01100111 01100111 01100011 01100011 01100111 01100011 01100111 01100001 01100001 01100111 01100011 01100111 01100001 01100011 01100011

In the case of cells, the information is stored in DNA. Rather than 0s and 1s, it is a string of As, Ts, Gs, and Cs. The order of nucleotides in DNA stores the language that cells know how to read and write.

GGCGAAAACGAAACCATTTGCGAAAACGGCATTAACGAAGAACGCATCGGCATTAGCGGCCGCGAAGCGACC

Is DNA the chicken or the egg? *Breakout Discussion*

Have you ever wondered what came first, the chicken or the egg? Now you can take that contemplation one level deeper!

If DNA is the blueprint for a cell, and cell machinery is required to read DNA, create DNA, and also to produce machinery of the cell, where did the first cell or DNA come from? What came first, DNA or cell machinery?

If there was no cell, how was DNA created and read? If there was no DNA, where did a cell and the cell machinery come from?

Ever since scientists identified that DNA is the material of inheritance, this question has persisted. Present-day theory suggests that neither DNA or the cell was first! Another molecule that will be covered in Chapter 4, called ribonucleic acid (RNA), is presently thought to have been the first active molecule. RNA, like DNA, is like a blueprint for cell machinery but is also able to cause chemical reactions to happen, like those performed by cell machinery.

But this is just a theory. We still do not know the answer...yet!

Summary and What's Next?

We did our first experiment where you suspended fruit cells in a salt slurry, lysed them open with a surfactant to release the DNA into the environment, which you filtered and precipitated. Were you surprised at the amount of DNA in a strawberry?

The hands-on skills and underlying chemical principles you learned during the exercise are essential and broadly used in genetic engineering. In fact, you will be using them in later chapters, so try not to forget them!

In this chapter, we also began to look at the structure of DNA to better understand why you were able to purify and precipitate the DNA using the chemistry you employed. We then discussed how DNA is like the blueprint of cells, holding information about how cells make important cellular molecules, and how to grow and divide. When we looked at DNA up close, we saw that there were four nucleotides named after their nitrogenous bases (adenine (A), thymine (T),

guanine (G), and cytosine (C)) that are all made of CHOPN atoms. The nucleotides are strung together into strands by cell machinery. It is the order of the nucleotides (the sequence) that holds the information that cell machinery understands how to read and write. You too will know how to read DNA by the end of Chapter 4!

We learned about how DNA can be changed and modified by humans, either randomly or very precisely, and that we're going to use these precise methods to change the cell's genome in later chapters. More broadly, we learned about how DNA changes naturally in the environment and how those changes are at the core of evolution.

In Chapter 2 we are going to look at what it takes to set up a safe genetic engineering area at home, school, and the makerspace, with a Minilab and explore biosafety considerations so that you will be ready to become a Genetic Engineering Hero!

Review Questions

Hands-on Exercise

1. What does a surfactant do?

2. Why do you break the strawberry apart into separate cells

3. What is precipitation? Find another analogy to explain precipitation.

4. Why did you use a filter during the exercise?

5. Why was the filtered liquid slightly red/pink (if you used a strawberry)?

6. Explain the chemical property differences between water and isopropyl alcohol.

7. What is a chelator? Why is one used in this exercise?

Fundamentals

1. What are the four different macromolecules?

2. What does DNA stand for?

3. Is DNA a permanent molecule? Explain.

4. What are the four nitrogenous base names?

5. What is the difference between a chromosome and a plasmid?

6. Who is Chargaff and what did he discover?

7. What is the DNA backbone? What is it made of?

8. What is a gene?

Chapter 2

Setting Up Your Genetic Engineering Hero Space

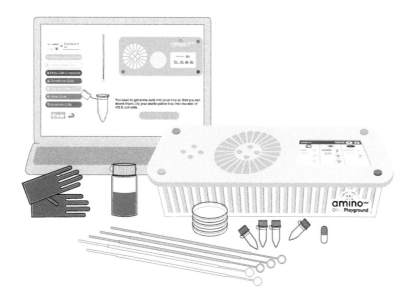

The world is entering a new, exciting era: a Biorevolution is happening! Until a few years ago, it was almost impossible for someone without years of university or professional training to do genetic engineering projects at home, in elementary and secondary school, or community spaces. Learning the fundamentals of genetic engineering, then applying them to do world-changing projects was possible only to those at universities, large companies, or in the government.

Thanks to advancements in technological and scientific innovations in hardware, software, and wetware, as well as in manufacturing, Minilabs, a type of laboratory equipment made for everyone to use and learn with, are revolutionizing this field, making genetic engineering accessible to those curious about the power of the genome and programming biology.

In this chapter, we will look at what Minilabs are and how to use them in your journey to becoming a safe and responsible Genetic Engineering Hero. You will

not need a large space to make into your lab space, nor will you need large and expensive equipment. Instead, you will use compact, simple-to-use lab equipment that can fit on your desktop. If you wanted to, you could even bring everything you need to a friend's house or a school.

We will start by talking about your workspace, the type of environment you should have for your experiments, and how to make sure you are following your local rules and safety guidelines. We will then look at the equipment and other materials you should have on hand to keep your workspace operating smoothly and safely.

If you are not completing the hands-on experiments while reading this book, not a problem! You can have a look through this chapter to learn about biosafety, governmental regulations and considerations to take when doing lab work.

Getting Started

Do I need government approval?

For the exercises included in this book, you will learn by experimenting with microorganisms that are non-pathogenic (not harmful). These organisms are classified as Risk Group 1 (RG-1), also called Biosafety Level 1 (BSL-1), precisely because they are not harmful. Note that this chapter provides suggestions for setting up an RG-1 ready space where you will be able to do all of the experiments in this book. For more advanced experiments outside this book's scope, such as those using organisms classified Risk Group 2 (RG-2) or higher, you would require a specialized lab space with governmental approval no matter where you are in the world.

However, because we will be learning with RG-1 organisms like most university students and researchers, you are likely not required by law to do your genetic engineering learning in a government-approved space. This does depend on where you live, as each country has different regulations. For example:

- In North America, you are free to do all the genetic engineering we discuss in this book at home, in schools, and makerspaces, without needing governmental approval.

- If you live in Asia and Africa, you are likely free to learn and complete RG-1 genetic engineering in the type of space we help you set up in this chapter. However, you should check with your local government to be sure.

- If you live in Europe and South America, there is a high chance that your government will want you to learn genetic engineering only in spaces that they approve and are aware of.

One of the advantages of the Minilabs used in the Genetic Engneering Hero's learning journey is that they are easy to use and to move around. If you do live in a country that requires genetic engineering to be completed in a government-approved lab, you can bring your Minilab there to do the learning exercises!

The rest of this chapter assumes that you can set up space in your home, school or community/maker space to complete the RG-1 level experiments. If you cannot conduct the experiments at home because of your country's laws or other reasons, you can still read on to learn what equipment you will need at a government-approved community or school lab.

Contacting your local government *Going Deeper* **2-1**

If you do not live in North America, you should search the internet to see if your country requires you to get approval or a license for completing basic genetic engineering experiments. Search terms like *"Biosafety Level 1 experiment in [country]"* or *"Risk Group 1 (RG-1) experiment in [country]"* should get you part way there, along with information about the government organization that you can contact directly.

By using the contact link on the government organization website, you can then send this email:

Hello,

I am interested in learning the basics of genetic engineering. I am aware that in North America I do not need government approval for risk group level 1 (RG-1) experiments. In [your country], am I free to complete RG-1 genetic engineering experiments with standard non-pathogenic lab strains of E. coli at [home/classroom/makerspace]? Do I need to get approval, a license, or have a certified space? If so, what is the procedure for doing this?

Thank you,
[Your name]

Figure 2-1. Set up your Genetic Engineering Hero headquarters in a safe, mold-free and insect-free environment.

What type of room should I set up my genetic engineering space in?

You can set up your genetic engineering space in most rooms. However, avoid the kitchen and bathroom. You should not set up your space in the kitchen because it is good practice to always keep your experiments separated from food storage, preparation, and eating. You should avoid the bathroom because there are often molds and other bacteria that love to grow in the damp environment. While they are unlikely to harm you, they will quickly show up to contaminate your experiments. Similarly, you should avoid damp or moldy rooms for the same reasons. Here are considerations when choosing your space:

- **Temperature:** When performing experiments, the temperature of your space should be between 17 °C and 25 °C. This range is approved for Minilab use and is also important for handling samples. Your space should ideally be in this range at all times. If your kits, and in process experiments are stored at the temperatures recommeded in the manuals (usually in a refrigerator), then the temperature of your space can be out of the recommended range when you are not using it. Once you are ready to work, we recommend using a heater or air conditioner to get your space within the ideal temperature range. The ideal temperature to aim for is 21 °C.

- **Free of insects and rodents:** Insects and rodents are pests and vectors (transmitters) for microbes. These could crawl into your samples or like their taste, so don't let your reagents (a substance or mixture used in an experiment) and other supplies get contaminated!

- **Good air quality:** When doing most of the projects in this book, you will be growing bacteria on nutrient-rich media (a solid, liquid or semi-solid made to support the growth of microorganisms or cells). Molds floating in the air love to feed on nutrient-rich media. If you notice some strange 'stuff' growing in your petri dishes, then you should definitely consider getting a small air purifier with a high-efficiency particulate air (HEPA) filter. Don't worry, you will learn to distinguish strange "stuff" *vs.* your experiment in later chapters.

- **Working surface and floor should be sturdy and easy to clean:** Setting up on a hard, non-porous surface made of metal, ceramics, or plastic is ideal. Whether you're working on a desk, in a closet or on a countertop in a classroom, as long as the surface will not tip over and can be cleaned by wiping, you are good to go. Because we are working with Minilabs, you won't need more than one square meter of space (3 x 3 feet) to perform your experiments. The floor surrounding your working surface should also be solid and non-porous so that it can be wiped and disinfected in the event of a spill. Carpet is not ideal because you cannot wipe it; if you need to set up in a carpeted space, you should acquire a rubber or plastic floor mat to set under your work area.

- **Cold storage (fridge/freezer):** As you progress through this book, you will be using kits that should be refrigerated. While it is OK to put the new kits in your refrigerator as long as the package isn't opened, we still recommend putting all your science materials in a sealed plastic container that you can keep in the refrigerator. Some experiment kits will have components that will need to be kept frozen and you can use a standard freezer for this.

While you are in the middle of your experiments or once they are completed, it is not recommended to put petri dishes containing grown or engineered cells directly in your fridge. While the cells you use in these experiments are not pathogenic, it is still recommended that you follow the golden rule of separating food from experiments! A solution is to use a large airtight Tupperware-type container that can seal your experiments when you store them in the fridge to prevent them from coming into contact with food. Always advise your family or roommates if you have science in the fridge!

Once you progress towards becoming a Genetic Engineering Hero, it is recommended that you acquire dedicated a mini-fridge, a thoroughly cleaned used fridge, or a refrigerated cooler to keep your experimental samples cool for longer storage. Consider this as soon as you can.

- **Pets or small children:** Just as you want to maintain a pest-free space, you need to prevent pets and children from touching or eating your experiments! To help you, the Minilabs have built-in locks for locking in samples while incubating. However, you may need a locked room or cabinet to store the rest of your experimental samples and to keep curious pets or children from interferring.

Setting up your space should be straightforward. We have completed experiments in many different environments and owe this to the Minilab; we've set it up in state-of-the-art university teaching labs at the Massachusetts Institute of Technology, in budget hotel rooms, and even in the car on longer road-trips. Consider the simple guidelines above, and get going!

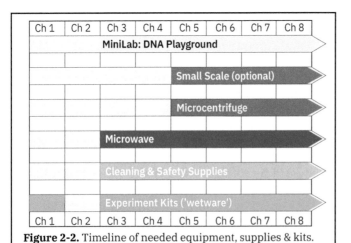

Ch 1	Ch 2	Ch 3	Ch 4	Ch 5	Ch 6	Ch 7	Ch 8
MiniLab: DNA Playground							
				Small Scale (optional)			
				Microcentrifuge			
		Microwave					
		Cleaning & Safety Supplies					
		Experiment Kits ('wetware')					
Ch 1	Ch 2	Ch 3	Ch 4	Ch 5	Ch 6	Ch 7	Ch 8

Figure 2-2. Timeline of needed equipment, supplies & kits.

Equipment and materials for your Genetic Engineering Hero space

Setting up your space is almost as simple as setting up a new computer or home office! In Figure 2-2, you can see a timeline of the equipment and materials you'll need for each chapter if you are doing all of the experiments in this book. Let's start by looking at the equipment that you'll need to complete the full journey: a Minilab, a microwave, microcentrifuge and a small scale.

1. The DNA Playground Minilab

Incubating, cooling, heatshocking and transforming samples (Chapter 3-7)

Figure 2-3. The DNA Playground Minilab and an experiment kit set up on a home desk covered in epoxy for easy cleaning.

The DNA Playground is about the size of a toaster oven. It will enable you to incubate and engineer cells. This versatile Minilab replaces the need for a large petri dish incubator, an ice bucket with a constant supply of crushed ice, a hot water bath, and a thermometer. This Minilab is used in chapters three through seven and will be essential for all of the exercises in this book!

The DNA Playground is available in two sizes: Home (Small) and Classroom (Large). There are two differences between them:

- The incubator capacity: The small unit can incubate two of the 6cm petri dishes at a time. The large unit can incubate eight of the 6cm petri dishes or two large 10cm petri dishes at once. In the kits recommended in this book, you will only be using 6cm petri dishes. If you will be working with a partner and will each be doing your own experiments, then you may want to consider the Classroom (Large) size DNA Playground since you can have up to four experiments running simultaneously!

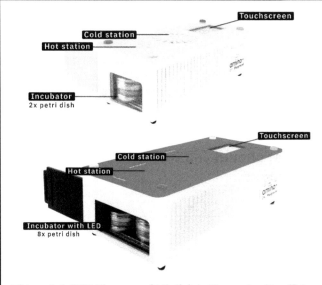

Figure 2-4. DNA Playground Minilab in Home size (Small) in yellow (top) and Classroom size (Large) in red (bottom).

When you do more experiments in the future, you may encounter the larger petri dish sizes. Amino Labs provides small petri dishes as they save materials and cost for you, and are readily available.

- The large DNA Playground comes with a built-in LED (light) for use when completing one of the Chapter 7 experiments, the Light-it kit. If you have the small DNA Playground, you can get the Light-it LED and a power pack for it when you purchase the Light-it Kit. Note that for Chapter 7, you have three experiment kits to choose from so you can also decide not to do the Light-it Kit.

Cost: $489-539 at Amino Labs https://amino.bio. Available individually or as part of the *Zero to Genetic Engineering Hero Pack Ch. 1-4*. Use code *GEH2BOOK5* for 5% off.

2. The Microcentrifuge

Extracting cultured products (Chapter 5-6)

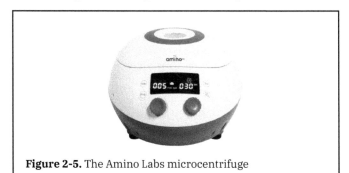

Figure 2-5. The Amino Labs microcentrifuge

A microcentrifuge is a compact version of a centrifuge that is about the size of a waffle iron. You will use this centrifuge when you begin breaking open cells to extract out the products that you engineered the cells to create. The microcentrifuge is required for Chapter 5 and 6.

Your microcentrifuge must meet two requirement for it to work with Amino Labs kits:

- Spin at 13,000 x g (~13,000 rpm) or higher
- Hold 1.5 mL centrifuge tubes

Cost: $500-$800. Get a new microcentrifuge at Amino Labs https://amino.bio. Available on its own or as part of the *Zero to Genetic Engineering Hero Pack Ch .5-7* or look for a used one on eBay, Amazon, etc. You can use discount code *GEH2BOOK5* to get 5% off at Amino Labs.

3. The Microwave

Creating Agar petri dishes (Chapter 3-7)

Figure 2-6. A Microwave Oven to boil your sterile water. *Copyright Consumers Union of US., Inc. 2015 All Rights Reserved.*

A microwave will be used to heat sterile water to a boil. This is a better tool for your lab than a kettle since you will be able to keep your sterile water in its sterile bottle in the microwave, preventing any contamination. Because no cells or microorganisms will ever be inside, you can use your existing kitchen microwave or get a new/used one specifically for your space. The water you will heat up will be more sterile than your microwave and tap water would be! It would help if you spent a little time cleaning the microwave before using it not to contaminate your experiments. Be careful when you heat water as it will be very hot to the touch!

The size and power of the microwave are flexible, and you will find the right settings to boil water as you go. The microwave is required for Chapter 3 and beyond.

Cost: Variable, likely less than $100. Look in your local second hand stores, or on used goods websites.

3. A small scale

Weight your samples for centrifuging (Chapter 5-7)

Figure 2-7. A small scale is used to balance samples before centrifuging.

To complete your lab, you need to get a small scale that has two or more decimal places when measuring in grams (0.00 g). Your scale only needs to be able to measure up to 300g. You will use this scale for weighing samples before you centrifuge them to ensure they are balanced. It is required for Chapter 5 and beyond.

Cost: Variable, but typically less than $50. Have a look online - Amazon and eBay are good places to start.

That's all the equipment you will need in your lab space to become a Genetic Engineering Hero. Isn't it amazing that you can do university-level experiments with so little equipment?!

But make sure to continue reading to learn about the safety equipment and items you will need. A Genetic Engineering Hero always practices safe science.

Materials and Supplies

In addition to equipment, you will need a few more items so that you can complete the experiments professionally and safely. These will also ensure that you are keeping your space running smoothly and are minimizing contamination.

1. What should you wear? Your Safety equipment

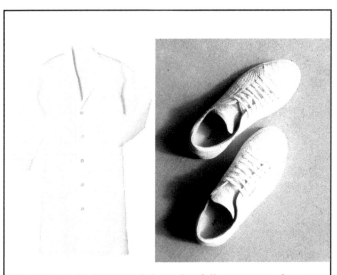

Figure 2-8. A lab coat and shoes that fully cover your feet are standard lab apparel.

A lab coat or apron: A lab coat or plastic (wipeable) apron will prevent contaminants on your clothes from falling into your lab space and experiments.

Do I need a microscope? *Going Deeper* **2-2**

In the world of genetic engineering, synthetic biology, and molecular biology, you work at the molecular level (nano or pico levels) rather than the micro-level. The molecular level is about 1000x smaller than what can be seen by a standard microscope. While you do use bacteria, genetic engineers generally do not need to look at the cells. Genetic engineering really is nanotechnology.

Microscopes are sometimes, but not very often, used by genetic engineers. Most of the time, analyzing and understanding the chemistry in the cells is the most important factor, and chemistry isn't easily seen using a standard microscope. Instead, other analytical tools are used to understand what is happening at the scale of atoms and molecules.

There are "microscopes" that can be used to analyze cells at the level of molecules, but even though they are called microscopes, they are actually "nanoscopes" that are capable of seeing details 1000-100,000x times smaller than the typical microscope. These devices include scanning electron microscopes, transmission electron microscopes, atomic force microscopes, and scanning electrochemical microscopes.

Contaminants can be your hair, your pet's fur, pollen, dust, food... all things you do not want getting into your work! Your lab coat will also prevent you from spilling your experiments on your clothes. Some of the cleaning agents you will use in your workspace are chlorinated and, by wearing a lab coat, you will also protect your clothes from getting bleached. You can find a lab coat online (e.g., Amazon, eBay, Amino Labs) for $30 or less. You can also use an artist smock or similar. If you get one of the *Zero to Genetic Engineering Hero Pack Ch. 1-4* at Amino Labs, you can add a mini safety set with a reusable apron you can easily clean.

Shoes: During your experiments, you should wear shoes that fully cover and protect your feet. No open toes or socks! You want to cover your feet just in case you spill hot or caustic materials - they will protect your feet. If house rules about wearing shoes in the house or community space exist, sturdy and fitted leather or rubber-type slippers can do.

Safety glasses / Eye protection: Eye protection like safety glasses can be worn throughout the experiments but are not always necessary since the quantity of the reagents is minimal, and the samples are safe. If you are not familiar with working in a lab, you may not be used to being aware of what you are doing with your hands... Do you scratch your face, rub your eyes, push hair off your forehead? Wearing safety glasses can help you prevent these unconscious gestures during experiments. Safety glasses should be worn when handling bleach in the inactivation step and in Chapter 5 and 6 for the filter-sterilization.

Cleaning and Other Supplies

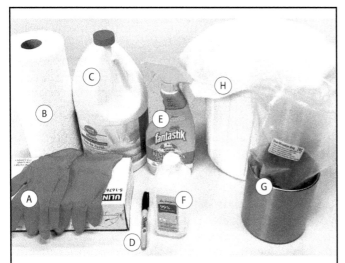

Figure 2-9. Your workspace should be stocked with some simple and useful items.

Lab gloves (A): It is recommended that lab gloves be worn at all times. Gloves will protect you from your experiment and your experiment from you. Nitrile or latex gloves are widely used and can be found at your local pharmacy or purchased online. A standard box of 100 gloves will be enough to complete all of the experiments in this book.

Paper towel (B): A roll of paper towels is great to have nearby in your space. Use paper towels to wipe up any spills. To clean up a spill, directly spray the spill with a generous amount of chlorinated cleanser and place a paper towel on top. Leave it there for 30 minutes, then toss it out.

Concentrated liquid bleach (C): Bleach is used to "inactivate" and kill bioengineered organisms after you are done with them and clean up your space and manage any spills.

Permanent markers (D): Labeling your samples, tubes, and plates is an integral part of doing genetic engineering projects. Labelling is important for safety and for organizing your experiments.

Spray Cleaner (E): A chlorinated cleaning agent should be used as your primary cleaner to spray onto a spill. Lysol or Fantastik spray with bleach are good examples. Any spray cleaner that includes sodium hypochlorite in the ingredients list is good. Be careful, since these can "bleach" clothing and other materials.

Alcohol Cleaner (F): Isopropyl alcohol (rubbing alcohol) or ethanol (sometimes called rubbing compound) can be useful to have around in either spray or bottle form. It is common to spray your hands with either ~70% ethanol or ~70% isopropyl alcohol after putting gloves on to provide a little more sterility before starting your experiments. Either of these can also be used to further sterilize your workspace after using chlorinated bleach spray cleaner.

Discard container (G): A small pail is excellent for placing inactivation bags into ("inactivation bags" are included in many of the kits you'll use throughout this book). Any experimental waste (items that have touched DNA, cells or reagents) will go into an inactivation bag. This waste includes gloves that have touched bacteria, used paper towels, tubes, inoculation loops, and more. In Figure 2-7, an inactivation bag has already been set into the discard container.

Garbage pail (H): You will also need a 10-20 litre plastic garbage pail for general garbage. Make sure the pail does not have any hole so it can hold liquid and line it with a "heavy duty" plastic/garbage bag. Any kit packaging can go into this waste container, as well as gloves that have not been contaminated with bacteria. In the very rare instance where a lot of material gets contaminated from an experiment, you can also use this garbage pail to dispose of your experiment waste. You will then be able to add bleach and water to the pail, inactivating the samples inside, because it is waterproof and lined with a sturdy garbage bag.

Experiment kits/wetware

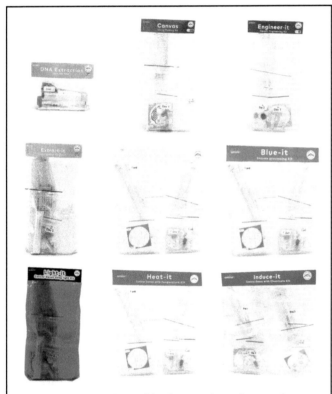

Figure 2-10. Experiment kits from Amino Labs are a fun and easy way to have every 'ingredient' you need to learn biotechnology hands-on!

For the rest of this book, the hands-on experiments require specific Experiment ('wetware') kits created by Amino Labs. You can get these kits as-needed by visiting https://amino.bio or get them all as part of one of the *Zero to Genetic Engineering Hero* bundles.

The *Zero to Genetic Engineering Hero* packs come in different options:

- By book chapters: *Chapters 1 to 4* and *Chapters 5 to 7* include the kits needed to complete the specified set of chapters. The *Ch. 1 to 4* pack includes the option to add a DNA playground of your choice

to the bundle, as well as the book (which you probably won't need if you are reading this already!) The *Ch. 5 to 7* pack includes the option to add the microcentrifuge and the Light-it LED add-on if you don't have the DNA Playground Large which has the built-it LED.

- By size variants - Home or Classroom pack. This is because Amino Labs experiment kits are available in two sizes. In individual size, you get one experiment 'try' per kit, and in Classroom size you will get 4 or 8 'tries' of the experiment per kit. The kits from *Ch. 1 to 4* Classroom size come in 8 tries, and the more advanced kits of *Ch. 5 to 7* Classroom size come in 4 tries.

You may be wondering - a Chapter 1 kit? But I've already done that! If you've already completed the DNA extraction from strawberry, you can buy the kits for the other chapters separately or get the bundle and do it again. After all, practice makes perfect!

Below, and at the beginning of each chapter, you will find a list of the kits you will need.

- **Chapter 1:** DNA Extraction Kit (or home materials)
- **Chapter 2:** No kit!
- **Chapter 3:** the Canvas Kit to make living paintings
- **Chapter 4:** the Engineer-it Kit to engineer bacteria
- **Chapter 5:** the Plate Extract-it Kit
- **Chapter 6:** the Smell-it Kit and/or the Blue-it Kit
- **Chapter 7:** the Light-it Kit and/or the Heat-it Kit and/or the Induce-it Kit.

Safety and Best Practices

There are some basic safety considerations and best practices that you should use when doing any biological experimentation. These guidelines will help to ensure that your workspace is a safe environment for you and for those around you. They will also set you on track to work or study in a university or company lab.

Some considerations were mentioned earlier in the Materials and Supplies section, and this section expands on this topic.

Do not eat or drink near your experiments or in your lab space. Keep your experiment at least 10 feet from food, drinks, *etc*. Under no circumstances should you consume any of the ingredients. Protect your experiments so that they do not get eaten by pets, pests, siblings or anyone/anything else!

- **Immunocompromised persons:** While components of BSL-1/RG-1 experiments like the ones for this book are non-pathogenic, some individuals, such as those who are immunocompromised, can be affected by large numbers of bacteria and should wear extra protection, such as long sleeves and a face mask, to ensure no contact with any experiment component. If you or someone around you has a compromised immune system, you or the immunocompromised person should consult a physician about the activities before starting any experiment.

- **Wash your hands** before and after manipulating your experiment, the ingredients, or the hardware.

- **Wear gloves** even when cleaning your space or handling the consumables (petri plates, loops, *etc*). This will protect you from your experiment, and your experiment from you. Any latex, nitrile, or general-purpose gloves you find at the pharmacy will do. Also, after you put on your gloves, be aware of what you touch. Do not touch your face or scratch itches with your gloved fingers as this will transfer microbes from your body to your hands!

- **Clean up your station, spills and work surface before and after use**. Use an alcohol solution (*e.g.* 70% rubbing alcohol), a chlorinated spray cleaner, or a 10% solution of chlorinated bleach generously applied to a paper towel and rub onto any contaminated surfaces. You can also use chlorinated wipes to do this. But be careful! These cleaners can discolor your clothing. Even if you have not had a spill, wipe down your workspace table with a 70% rubbing alcohol solution before or after your work but never use rubbing alcohol on your minilabs. To clean minilabs, refer to their instruction manuals.

- **Discard used consumables properly.** During your experiments, you will have ingredients, consumables, and microorganisms that you need to dispose of safely. A simple inactivation procedure is described for each of the hands-on exercises. This includes putting any used items inside the inactivation bag (which has been placed within your discard container). Then add bleach and water to the inactivation bag. After 24 hours, all organisms will be killed, and DNA will be denatured (broken down and made non-functional). It is safe at that time to put the liquid in the toilet and dispose of the solid materials in the garbage.

- **Tie up your hair.** If you have long hair, be sure to tie it up in your lab space. Hair can get into an experiment and contaminate it, while also contaminating your hair with the experiment components.

Print a few **Safety and Clean-up checklist** and keep them by your workspace to use before and after your experiments as you become familiar with the procedures. https://amino.bio/checklist

The Public Health Agency of Canada (PHAC) also has some excellent materials that expand on this topic:

Biosafety 101 course: A free online biosafety course is offered by PHAC. It covers the biosafety basics that you should consider, especially when working in a professional lab. You will be redirected to this resource by visiting https://amino.bio/biosafety.

What-is-Biosafety? Poster: A biosafety poster that describes what biosafety is all about. It will teach and remind you that your eyes, mouth, cuts and nose are your main vulnerabilities to the world around you. Download and print this poster to put it up on your Genetic Engineering Hero lab wall. You can find this resource at https://amino.bio/whatisbiosafety.

Biosafety-in-Action Poster: A biosafety poster that you can print and put on the wall of your space. This poster describes actions you should emulate while at your genetic engineering workspace. You can find this resource at https://amino.bio/biosafetyinaction.

Biosecurity Poster: By completing the exercises in this book, you will gain the superpower of genetic engineering. You will be an essential player in making sure that you and no one else uses genetic engineering to do bad. Have a look back to the beginning of the book, recite the Genetic Engineers' Pledge and sign it.

PHAC created a biosecurity poster so you can learn what biosecurity is and the importance of safe genetic engineering. While this poster mentions "bacteria" as a biosecurity risk, remember the difference between "Risk Groups". You are only using RG-1 organisms throughout this book. The biosecurity poster, in general, refers to "pathogenic organisms" that are RG-2 or higher. Find the poster at https://amino.bio/biosecurity.

Just because I Can, does it mean I should?

Technology is like a double-edged sword. This means that while technology has the potential to do immense good and even propel humanity to the next phase of its evolution, technology can also do harm at the level of the individual, environment, or even populations. Biotechnology is no different than the great technologies that preceded it.

For example, technology that enabled the reliable creation of fire kept our ancestors warm during winters, even entire ice-ages. However, even today, fire is the cause of millions of deaths every year due to smoke inhalation and carbon monoxide poisoning. One could even argue that harnessing fire in the form of the internal combustion engine has led to countless more deaths every year due to car accidents and the unprecedented creation of greenhouse gases like carbon dioxide, which has the potential to alter global climate.

Ethical review boards are a common tool used by universities, corporations, governments, and non-government organizations to screen for negative impacts of technology development. A long-standing dilemma is the use of animals during research studies. Today, stringent rules are in place for animal use, and alternative methods of study must be used if possible. In other words, you are smart, and there are probably other methods or experiments you can use toward building your technology. Even if your desired technology is new brain implant technology where you require animal(s) to test your system, you can easily design methods and experiments to test out your technology using neuron cells in Petri dishes. Only after you have demonstrated that your technology works and will not kill neurons in a petri dish should you even consider using animals.

There is much pain in our world, and you want to bring positive change, so it is your duty to use your ingenuity to minimize the creation of negative impacts of your experiments and technologies. Unforeseen circumstances of the development and deployment of technology are often difficult to predict. Still, with dedicated contemplation, you can do your duty as a responsible scientist/engineer/citizen to help bring about positive change.

Always keep in mind that the conversation about the technologies you will create during your lifetime is ongoing and evolving - you should always be asking the question: Just because I can, does it mean I should?

Should I? *The 3 Levels of Technology Contemplation Breakout Exercise*

While there is no perfect tool to assess for unintended consequences or whether your technology is ethical, try using Amino Labs' *Three Levels of Technology Contemplation* to assess any technology projects you start. Don't forget, you can share your work with the Genetic Engineering Hero community (amino.bio/community), and you might get some feedback too!

Technology Name:
Intended impact:

For this technology, List two unintended secondary impacts that could occur if your technology becomes a normal part of society. You should list one positive impact and one negative impact:	
Positive Impact	**Negative Impact**
A1.	A2.

List two unintended tertiary (third level) impacts that could occur if your technology becomes a normal part of society. You should list one positive impact and one negative impacts for each of your above listed secondary unintended impacts:	
Positive Impact	**Negative Impact**
A1.	A1.
A2.	A2.

Thinking about science... in fiction! *Breakout Exercise*

One way technologists and engineers think about the future is by looking at the potential social impacts that imagined or emerging technologies could have... in science fiction! Whether in film, TV, novel, video game or comic book form, most science fiction works expose a new world-view of society resulting from the emergence and adoption of a new technology or science. The example we are most familiar with in biotechnology is, of course, Jurassic Park. But, there are many more authors and creators who have crafted intriguing worlds borne from the impact of biotechnology. While many may portray negative views of such a future (dystopia), we technologists and scientists can learn a lot by using our 'ethical muscle' at the same time as we entertain ourselves.

Let's try! Go ahead and practice your 'ethical muscle' by reading the short story *Double Spiral* by Marcy Kelly, available to read online at Slate.com as part of their *Future Tense Fiction* series or at amino.bio/Doublespiral.

While you read or after you've finished reading *Double Spiral*, fill out Amino Labs' *The 3 Levels of Technology Contemplation* table using the primary technology described in the story. This is a technology you'll recognize from the real world, taken to some unintended places. (Hint: it's a saliva-ting technology!) You can use implied or obvious impacts to fill your table.

Then, add one more level of contemplation starting at the end of the story. Using your own imagination, think about where the technology in *Double Spiral* could go next? What would be the implications? For example, you could imagine what would happen if the protagonist did expose the alluded-to truth about *LyfeCode* and the *K5* mutation?

Since *Double Spiral* by Marcy Kelly is a dystopia, you will be exposed to a world in which *LyfeCode* technology brings about mostly negative impacts. You've identified a few positive impacts when completing *The 3 Levels of Technology Contemplation*. What are some other positive impacts that could of happen had the story been written differently? Try to re-write the story as a utopia (positive view of the technology and the future). What is different? What do you think could be done today to ensure we move towards a utopian end with this technology that is already part of our everyday?

If you enjoyed this exercise, have a look at Table 2 for more works that deals with genetic engineering or biotechnology. Of course, this is not a complete list but some selected works that we've found noteworthy.

Table 2-1 - Science fiction to explore biotechnology ethics	
Reading	
Maddadam trilogy by Margaret Atwood	Novel
Borne by Jeff Vandermeer	Novel
Beggars in spain by Nancy Kress	Novel
Always true to thee, in my fashion by Nancy Kress, (available online at Lightspeedmagazine.com)	Short Story
Brave new world by Aldous Huxley (A classic!)	Novel
Emergency Skin by N.K. Jemisin	Short Story
Flowers for Algernon by Daniel Keyes	Short Story
Blood Music by Greg Bear (short story version available online, see Freesfonline.net/authors/Greg_Bear)	Short Story & Novel
Watching	
Gattaca (1997)	Movie
Blade Runner (1982) and Blade Runner 2049 (2017)	Movie
Never let me go (2010 - based on a 2005 novel)	Movie/Novel
Okja (2017)	Movie

Summary and What's Next?

You now know the type of environment, equipment, safety clothes and other materials needed to complete all of the exercises necessary to become a Genetic Engineering Hero.

You've also learned that by becoming a Genetic Engineering Hero, you will have great power. Contemplation and ethical examination must always go hand-in-hand with great power. Always remember that just because you can, it doesn't mean you should.

You are now ready! Let's move forward into the world of genetic engineering bacteria by looking at what cells we are going to use throughout the book. In the next chapter, you are going to learn the skills for growing microorganisms.

Review Questions

Fundamentals

1. What does "BSL" and "RG" stand for?

2. Why is clean air important to have in your workspace?

3. What three things should you consider wearing during your experiments?

4. What seven best practices should you consider when genetic engineering?

5. Are you allowed to work with RG-2 organisms? Explain.

Chapter 3

Growing *E. coli* Cells

In Chapter 1, you learned about the structure of DNA and how it is a microscopic chemical string of nucleotides that is the blueprint for living cells. Cells know how to read DNA to build their structural components and internal machinery, as well as to create enough energy to grow and replicate.

In this chapter, we will focus on *Escherichia coli* (*E. coli*). *E. coli* is the microorganism that you will be engineering in the experiments in this book. Microorganisms are organisms that are not visible to the naked eye, like bacteria, viruses, and some fungi. *E. coli* is a type of bacteria from the *Escherichia* family. There are many types of *E. coli*, and all are living cells that contain a DNA blueprint.

The *E. coli* you will be using is not to be confused with the *E. coli* you may hear about in the news that can cause illnesses in humans or animals. The *E. coli* bacteria used in the Genetic Engineering Hero's journey comes from a safe lab strain of *E. coli* called K12. It is the most widely used organism in life science research labs around the world.

In this chapter, you will learn the skills and principles behind growing *E. coli* in a petri dish, an essential skill to master on your journey to becoming a Genetic Engineering Hero. You will make the "food" and surface (substrate) that *E. coli* grow on. You will learn about the environmental conditions that are

required to grow *E. coli*. You will learn the necessary techniques to grow cells so that they can be useful for other experiments.

Like Chapter 1, the Going Deeper sections provide context and more in-depth knowledge. They will give you a deeper understanding of what is happening throughout your experiment so that you can master the techniques and even modify them in future experiments. If this is your first time growing bacteria, you may want to repeat the exercise twice: Once with a focus on the hands-on instructions, and a second time focussing on the Going Deeper sections.

In the *Fundamentals* section, we will examine the "cells-as-factories" analogy, learning more about bacteria in general. In Chapter 1, we learned about the DNA as blueprints. In this chapter, we will take a tour of the entire *E. coli* microfactory, visiting the five key components of an *E. coli* cell:

- Capsule layer: outer protective shield of *E. coli*

- Outer membrane: primary outer structural barrier

- Intermembrane space: a key passageway to enter and exit the cell

- Inner membrane: an inner structural barrier

- Cytoplasm: the primary space in which the cellular activities occur

Keep in mind!

The *E. coli* you will use throughout this book is a specific lab strain that is safe. In fact, most types of *E. coli* are not infectious and are actually necessary for your digestive system to function. *E. coli* have evolved over millions of years inside the large intestine of animals such as humans. By colonizing our large intestines, they help us digest food, create vitamins, and produce amino acids for us. One strain of *E. coli* is a widely used probiotic that helps to reduce bloating in the intestine.

The strain of *E. coli* included in the Amino Labs' kits was first collected by scientists in the early 1900s. It has been used in thousands of labs worldwide and in countless experiments for nearly 100 years, with no reported incidences of harm. The *E. coli* within the kit also has further deficiencies, making it unlikely for it to survive outside your petri dish. You will learn more about this later on in the chapter. For now, remember that this bacteria is safe to use in your experiments. Even so, to get used to standard science safety procedures, you must always follow safety practices you learned in Chapter 2, no matter how safe the type of bacteria.

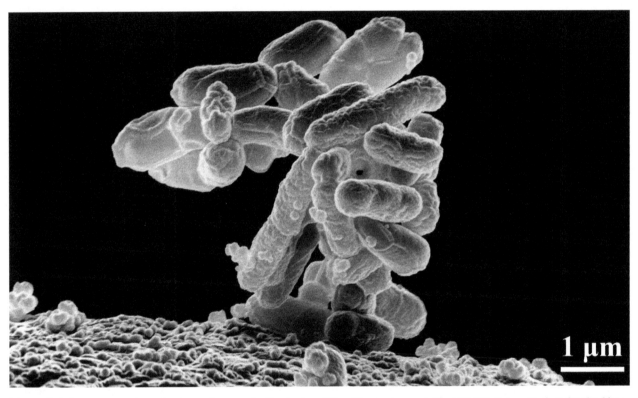

Figure 3-0. Low-temperature electron micrograph of a cluster of *E. coli* bacteria, magnified 10,000 times. Each individual bacterium is oblong shaped. 2005. Photo by Eric Erbe, digital colorization by Christopher Pooley, both of USDA, ARS, EMU., Public domain, via Wikimedia Commons

Getting Started
Equipment and Materials

For this chapter, you will need a **Canvas Kit™** from Amino Labs. The **Amino Labs' Canvas Kit™** is part of the ***Zero to Genetic Engineering Hero Kit Pack Ch. 1-4*** and includes all the required pre-measured ingredients. This kit can also be ordered separately at https://amino.bio/products.

You will also need a standard box of Jell-O™ for the Practice Exercise. The equipment you will need includes a DNA Playground Minilab, a microwave, and a computer (not a tablet or phone) to complete an online simulation.

Shopping List

Pre-practice exercise:
Box of Jell-O™ or Jelly™

Main exercise:
Wetware kit: Amino Labs Canvas Kit™ (https://amino.bio)
Wetware kit (optional): Keep-it Kit™ (https://amino.bio)
Minilab (DNA Playground)
Microwave
Laptop or Desktop PC with internet connection for completing Virtual Bioengineer™ Canvas Kit Edition.

Instructional Overview

A. Pre-practice by making Jell-O™

B. Pre-practice simulation using Virtual Bioengineer (https://amino.bio/vbioengineer)

1. Download the manufacturers Canvas Kit™ instructions from https://amino.bio/instuctions
2. Put on your gloves and lab coat
3. Creating molten LB agar
4. Adding antibiotics
5. Pouring LB agar plates
6. Use & storage of LB agar plates
7. Streaking and painting *E. coli* bacteria
8. Incubating *E. coli* cells
9. Viewing plates of grown bacteria
10. Preserving your bioart (optional - using Keep-it Kit™)
11. Clean up and inactivation

Chapter Timeline Overview

Timeline to complete the hands-on exercise is:

Day 1: ~20 minutes Breakout Session 1, ~20 minutes Breakout Session 2,
Day 2: ~60 minutes to make LB agar petri dish and streak cells, followed by 24-48 hours incubation
Day 3: ~30 minutes to paint your living paintings followed by 24-48 hours incubation
Day 4+: ~15 minutes to view results.

Timeline to read *Fundamentals* is typically 3 hours.

Making Jell-O™ *Breakout Session 1*

One of the first lab skills necessary to master in genetic engineering is the creation of LB agar plates. LB agar is a substance a bit like Jell-O™. You pour the agar as a hot liquid into a container – in this case, a petri dish (also known as a plate). The agar then cools to become the gelatinous substrate (or surface) on which the *E. coli* cells feed and grow. Since the principles of making LB agar plates are similar to making Jell-O™, that's where we'll start.

When making Jell-O™, you mix the powder with boiling water. It is important that this water is boiling because otherwise, the gelatin (gelling agent) in the Jell-O™ will not dissolve. Once everything is mixed together, the solution becomes transparent, with no particles are floating around.

The Jell-O™ instructions suggest you add cold water to accelerate the cooling process. You will not do this when making your LB agar plates but do follow your Jell-O™ instructions. Then, you will pour your Jell-O™ mix into a container and let the solution cool down and solidify. At the end, you have a transparent yet colorful gel. This is very similar to LB agar plates. Since you will be spreading (streaking) cells onto the surface of the agar in the main experiment, try running a spoon or chopstick on the surface of the solidified Jell-O™ to get acquainted with its texture and softness. This will help you work well with agar later on.

The key differences between Jell-O™ and agar are:

- Jell-O™ uses gelatin, a protein-based gelling agent from animals, whereas agar uses a sugar-based gelling agent from algae.

- LB agar has sugar, nutrients, amino acids, and minerals for bacterial growth - while Jell-O™, for the most part, has only sugar. Yum!

- Jell-O™ includes food dye to change its color, while LB agar is naturally a pale yellow due to the vitamins and other cellular nutrients.

- Jell-O™ requires you to add cold water and refrigerate it to speed up the "gelling" process. With LB agar plates, you do not have to do this because:

    a) the agar gels pretty quickly (10-15 minutes)

    b) it could compromise the sterility of your agar (by contaminating them with organisms in tap water).

    c) the antimicrobial chlorine and/or fluoride in tap water would prevent the growth of your bacteria.

Virtual Bioengineer™ simulation *Breakout Session 2*

The Virtual Bioengineer™ Canvas Kit Edition is an online simulation that mimics the procedure you will use to make LB agar plates. As you will see, making LB agar plates applies similar principles to making Jell-O™. However, the procedure and materials are different.

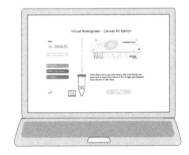

Virtual Bioengineer™ Canvas Kit Edition will familiarize you with the materials and procedure used in this chapter. The simulation is free and will take less than 20 minutes to complete. https://amino.bio/vbioengineer

Learning Hands-On: Growing K12 *E. coli* cells

Step 1. Download the instruction manual for the Canvas Kit

Figure 3-1. Step 1 Download the Canvas Kit instruction manual. Make sure to choose the right one from 'Individual or Classroom' based on the kit you have.

Familiarize yourself with the Canvas Kit instructions found at https://amino.bio/instructions.

The manufacturer's instruction manual has the most up to date procedures for this exercise and may be slightly different from this book. If there are any differences between this book and the up-to-date manufacturers' instructions, always follow the manufacturer's instructions.

Step 2. Put on your gloves and lab coat

When doing genetic engineering and science experiments, you should practice what is called aseptic technique. Aseptic technique is the use of methods and procedures to prevent cross-contamination of microorganisms between your experiments, your environment, and you. Much of the following has been talked about in Chapter 2, but let's quickly refresh the topic.

You have lots of microorganisms on your hands that can easily be transferred into the petri dishes you will be making in this exercise, and can result in contamination of your experiment. The same goes for your clothing, mold in your environment, your pets, and even your hair. Conversely, you will be growing microorganisms throughout this book, and you want to prevent them from getting onto your skin and body.

Practices like wearing nitrile or latex safety gloves at all times during experiments is a great way to prevent the transmission of microorganisms in either direction. Wearing a lab coat will also help prevent the transfer of microorganisms between the rest of your body and your experiments. If you have long hair, you should tie it up to prevent it from making contact with your experiments. While it is always recommended that safety glasses be worn, you will rarely see this as common practice in research labs. If you have a respiratory infection, you can wear a mask to prevent any microorganisms in your breath from getting in your experiments. Make sure to never touch your face with gloved hands. If you do, wash your face with soap and water and change your gloves.

It is also good practice to put the inactivation bag from your Canvas Kit into a discard container now so that it is ready for use and won't tip over.

Step 3.   Create molten LB agar powder

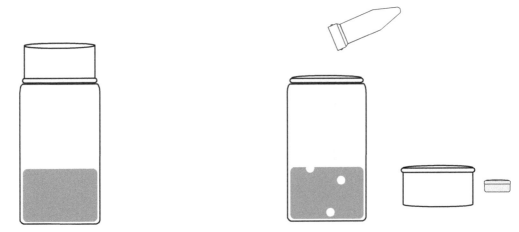

Figure 3-2. Step 3a: Boil your sterile water in the microwave.   **Figure 3-3.** Step 3b: Add the LB Agar powder to boiling water.

Get your bottle of sterile water and the tube of LB agar powder from your **"Day 1"** bag in your Canvas Kit and go to your microwave. Remove the lid of the **sterile water bottle** and set it loosely on top of the bottle. This prevents the boiling vapor from causing the bottle to explode. Heat the water in the microwave until it is boiling. A typical microwave will require about 45 seconds to boil the 50 mL bottle of water. Keep your eyes on your bottle and your finger on the stop button so that if you see it boiling over, you can press stop. You'll know you are ready to move onto the next step when you see bubbles actively bubbling up. If you don't see boiling, microwave in 5 seconds interval until you do see boiling.

Once the sterile water is boiling, remove it from the microwave and add the **LB agar powder** to the bottle. The bottle will be hot! Careful! A small amount of powder may stick to the lip of the LB agar tube, and that is OK and accounted for.

Once you have added the LB agar powder, partly screw on the lid and gently swirl to mix in the powder for 15 seconds. Swirling gently will help you avoid introducing bubbles in the agar. The powder may not fully dissolve because the water will have cooled down as the cold powder was added. Keep the lid loose and put the bottle back in the microwave. Heat in intervals of no more than 4 seconds. When it has boiled again, which you will see since it boils into a foam after LB agar is added, then take the bottle out of the microwave.

If the LB agar has boiled and is well mixed, it should look very clear with a yellow tint since LB agar is slightly yellow. If it looks like there are still some particles floating about, or if it looks cloudy, microwave for another 4 seconds to help dissolve the remaining powder. Be careful! the liquid boils over quickly once the LB agar powder has been added. Keep your finger on the button to stop the microwave if you see any liquid coming out of the bottle, or if the lid pops off due to bubbling. Move to the next step in less than one minute once your agar is boiled and gently swirl-mixed for 15 seconds.

LB agar *Going Deeper* 3-1

LB agar is a mixture of 'LB' and agar: LB is an abbreviation for Lysogeny Broth (sometimes also said to stand for Luria Broth, Lennox Broth or Luria-Bertani medium, after its inventor Guiseppi Bertani).

In all instances, LB refer to a "chicken noodle soup"-like mixture of sugars, salts, and minerals that *E. coli* bacteria love to eat. When in LB agar form, the agar is a gelling agent that will make the mixture solidify into a gel so that you can grow *E. coli* bacteria on its surface.

Step 4.   Adding antibiotic

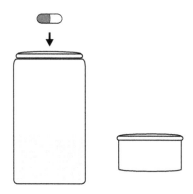

Figure 3-4. Step 4: Add the antibiotics to the LB Agar and mix by gently swirling (if applicable to your kit).

Antibiotics are frequently used in genetic engineering experiments. These allow you to "select" for the bacteria you want to grow while minimizing the chance for contamination. **Antibiotics** for selection will be discussed in Chapter 4. For now, follow the manufacturer's instructions on adding the antibiotics to your LB Agar. You must do this quickly before your LB agar sets.

Typically, these instructions will be along the lines of: Add the pill with antibiotics to your bottle of hot molten LB agar immediately, as soon as it has cooled down enough for you to hold the bottle for 5 seconds without burning your hands but before it gets cool. Swirl the bottle in a circular moti until the contents of the pill are dissolved. You may see some of the gel capsule floating around, and that is OK. You can move onto the next step.

If you see any of the powder that was inside the capsule remaining, you need to continue swirling. Try not re-heat the solution when mixing here because this will degrade the antibiotics. If you absolutely need to re-heat, go ahead, but know that your experiment might not work out perfectly.

Note that if you have the EU-compatible Canvas Kit with naturally colorful bacteria, you will not be adding antibiotics. You will learn why in Chapter 4.

Mixing *Pro-tip*

Mixing sounds easy and trivial, but mastering the art of mixing is a really important skill to learn on your journey to becoming a Genetic Engineering Hero. The first thing to remember is that you should always mix thoroughly when you add one tube of something to another, or if you add a powder into a liquid.

When mixing, it is important that you do not "half mix" something. Otherwise, your experiment will likely fail, or you will have inconsistent results. When mixing your agar powder into water, for example, you want to gently swirl so that the powder mixes into the liquid without any splashing that leaves powder stuck on the inside of the water bottle. In some instances, you may need to re-heat the sample to ensure that the powder has fully dissolved. In short, mixing is really important, and your goal is to achieve 100% perfect mixing without losing ingredients to splashing, spilling or evaporation. Scientists and genetic engineers get good at mixing over years of experiments, so be patient and be aware of this subtle but important skill.

Step 5. Pour LB agar plates

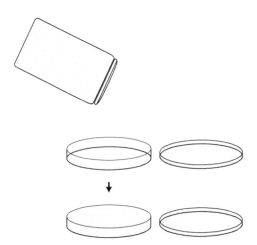

Figure 3-5. Step 5: Pour the LB Agar into the bottom half of the petri dishes (plates). The bottom is the side with the star-shaped border. The plates should be one half to two thirds full.

If your molten LB agar cools too much, it will solidify. If this happens, you can place it back in the microwave for 5-second increments until it is liquid again. Remember that heating with antibiotics in the LB agar can degrade the antibiotic!

Within the Canvas Kit, you will find **four petri dishes.** It is common in the scientific world to call these "plates."

When you get your plates, note that they are made of two parts: the top lid is slightly larger in diameter than the bottom dish and will overlap the bottom when you close it. During this step, and at all times when using plates, make sure you do not accidentally pour the LB agar into the plate lid!

Place the petri dishes on a clean flat surface, take the lids off, and place them next to the petri dish bottoms. The bottom is the half with the star-shaped border. Get your hot molten LB agar bottle, remove the lid and pour the molten LB agar into the dishes. Your goal is to fill the plates one half to two thirds full. Make sure the entire plate bottom is covered by the LB agar. Set the plate lids to cover three-quarters of the plate bottom, allowing heat to escape. Let the LB agar cool, solidify, and dry slightly. This usually takes between 15 and 30 minutes depending on the temperature and humidity of your workspace. Humidity and temperature play an important role in the timing of agar solidifying. For example, in the Canadian prairies during a blizzard, the humidity can be as low as 20%. This means that agar plates can solidify and dry in under 5 minutes! In a humid tropical location like Florida it can take 30 minutes or more! As noted in Chapter 2, if you are in an environment that might have mold, a HEPA filter is an excellent tool for minimizing contamination. If you think you'll have "dirty" air or have had some contamination in prior experiments, you can also fully cover the plates with their lids while your LB agar cools.

In Figure 3-6, you will see images of LB agar in petri dishes that demonstrate what fully dissolved LB agar should look like after it is poured in a petri dish and solidified. In Figure 3-6 (right), the powder was not fully dissolved, and you see particulates, which are making the agar opaque. If your LB agar powder was not fully dissolved, it will also be less solid and will be easily punctured with an inoculating loop. If this occurs, then the next time you make plates try microwaving your water-LB agar powder mixture for further 5-second intervals until it is boiling and fully dissolved.

At the end of this step, the manufacturer of the Canvas Kit suggests that you keep your sterile water bottle to measure and add bleach to the inactivation bag at the end of this exercise. You can rinse it out with tap water and screw the lid back on to save it for later. If you have leftover LB agar, you can pour it into your inactivation bag. If the agar has solidified, shake the bottle vigorously to dislodge the solidified agar and drop it in the inactivation bag. You can wash the bottle out with soap and keep it for future experiments or storage after you use it in the inactivation procedure.

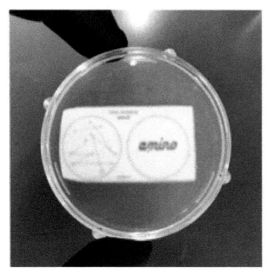 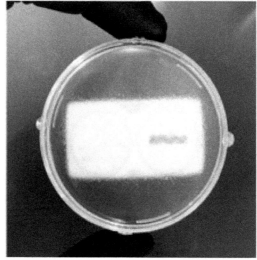

Figure 3-6. Fully dissolved LB agar powder in a plate (left) *vs.* incomplete dissolving of LB agar powder in a plate (right). Hold your finished plates 10 cm - 30 cm above a stencil to see whether you can see them clearly. Soft and semi-opaque agar means that you need to heat it more in future experiments. If this happened to you, don't worry! This is a part of your learning journey.

Sterile Water *Going Deeper* **3-2**

Sterile distilled water is provided within the Canvas Kit™. Sterile distilled water is used because it is purer than ordinary tap water. When doing scientific experimentation, it is essential to use sterile, pure ingredients and to do your best preventing environmental factors like spores, fungi, and mold from getting into or on your samples. Tap water and distilled water obtained in a store may contain bacteria. Tap water is generally OK for you to drink because your immune system will take care of any microorganisms living in the water. Your agar plates do not have an immune system, and any microorganisms in the tap water can and will grow in your experiments.

Research labs use an autoclave to sterilize water and other samples. An autoclave is a large and expensive pressure cooker that lets you heat the samples at, or above, 120 °C for long periods of time without the sample boiling away thanks to pressurization. For example, the sterile water you are using was autoclaved for more than 30 minutes at such a temperature.

As you become a Genetic Engineering Hero, you may want to complete your own independent experiments, and if you do not have an autoclave, you can boil distilled water that you get from a grocery store for at least 10 minutes.

Labeling Plates *Pro-tip*

Labeling LB agar plates is a very important practice. While it has not been mentioned yet for this exercise to maintain simplicity, in future chapters it will become common practice.

When you label your petri dishes in future experiments, use a permanent marker and write your initials on the **bottom** of the plate. It is essential to label plates on the bottom because lids can fall off or get mixed up during experiments. By labeling the plate bottom (where you poured your agar), the label will never get separated from your experiment.

In general, you should label the plate to indicate:

- If the LB agar has antibiotics, and if so, which antibiotic. Labs use color codes or abbreviations to identify the antibiotic in the plate quickly. Different labs may have different color codes. Table 3-1 identifies the color codes used by IGEM, the International Genetically Engineered Machines Competition. You can use it for your own Genetic Engineering Hero Headquarters.

- Your name or initials so that you or others will know the owner of the plate

- A descriptor to help tell you about the samples. For example, later on in the book, you will use S (+) to describe a positive control plate, and S(e) to describe your experimental sample.

Table 3-1 - IGEM Antibiotics Color Codes		
Antibiotics	**Color Code**	**Abbreviation**
Ampicillin	Orange	Amp
Chloramphenicol	Green	Clr/Chlor
Kanamycin	Red	Kan
Tetracycline	Yellow	Tet

Step 6. Use & storage of LB agar plates

Figure 3-7. Store your LB agar plates in the supplied resealable bag in the refrigerator for up to a few months. They can be used as long as no mold has grown on them. If you see mold, you must inactivate!

After the plates have solidified, you can immediately use them for your projects. You can also place your LB agar plates into the resealable bag they came in and store them for later use in a refrigerator or in a cool dark place. If the air in your environment is clean and no mold or spores have landed on your plates, then you can store your plates for a month or more in a refrigerator. If you have stored your plates but notice some unexpected growth on the LB agar, it is recommended that you inactivate those plates. Inactivation is covered in Step 13.

In the next step of this exercise, we will use one of the four plates you poured. Put three of them back in their resealable bag and refrigerate them. Once you have incubated the next step's plate for ~24 hours, you will use the remaining three plates to paint your living art.

Step 7. Streaking *E. coli* bacteria

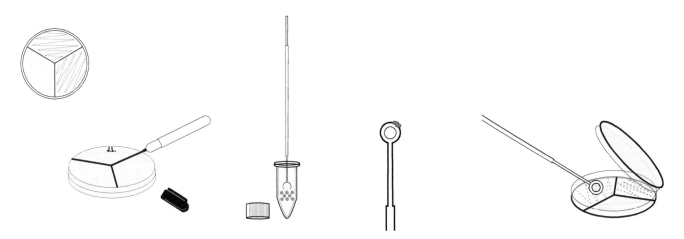

Figure 3-8. Step 6. Following the stencil example in the instruction manual, divide and streak one of your plates with the bacteria paints (colored bacteria) included in the kit.

Now that you have made your LB agar plates, you can grow bacteria on them. The Canvas Kit is meant for you to make bioart. By painting bacteria onto the LB agar using various utensils and then incubating them, you can create a living painting. As you learn to become a Genetic Engineering Hero, you will also practice a method of putting bacteria on an LB agar plate called streaking.

There are different types of streaking that you will learn as you follow the Zero to Genetic Engineering Hero Journey. Today's streaking is simple. Your goal is only to transfer some of the colored bacteria to your LB agar petri dish so that you have a greater quantity of cells (bacteria paint) that are freshly grown to use in your experiment.

On your one plate, you want to create a painting palette of bacteria paint that you will use as your "source" paint for your bio-art (living painting). To do this, you will streak out cells using a procedure that you learned and practiced in Virtual Bioengineer Canvas Kit Edition and that you can see step by step in the instruction manual:

1. See how many bacteria paint colors are included in your Canvas kit. This will typically be 3 colors. Using a marker, divide the bottom of your petri dish in the corresponding number of sections and label the sections with one section for each bacteria paint color you have.

2. Get a **yellow inoculating loop**. Dip it into one of the **bacteria paint tubes**. Inspect the loop to see if it looks "wet," indicating that you've dipped into the cells/agar. Then trace a zig-zag in the first section of your petri dish. Discard the loop in your inactivation bag.

3. Using a **new yellow loop** each time, repeat the above step for each tube of bacteria paint. Always discard your loops in your inactivation bag.

Step 8. Incubating *E. coli* cells

Turn on the **Incubator** of your DNA Playground Minilab to 37 °C. As the incubator heats up, you should flip your LB agar plates so that the bottom is up and the lid is down. Following the manufacturer's instructions, place the flipped LB agar plates with *E. coli* bacteria onto the paddle, cover with the humidity chamber, then place into the incubator. Turn on the timer to better monitor growth. Incubate the cells for 12 to 24 hours (but up to 48 hours if needed). You will start to see them after ~12 hours, and they will start to change color shortly after. You are now growing your first bacteria on your journey to becoming a Genetic Engineering Hero. Congratulations!

Step 9. Viewing plates of grown bacteria

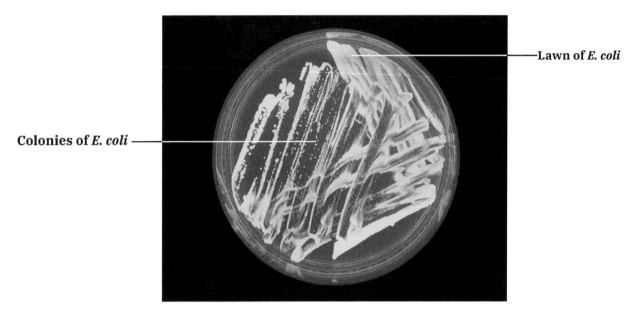

Lawn of *E. coli*

Colonies of *E. coli*

Figure 3-9. *E. coli* cells engineered to make cyan fluorescent pigment, streaked onto a selective LB agar plate, photographed under black-light (cyan bacteria will appear white under natural light). Characteristics of a successful experiment include: i) All bacteria on the plate are the expected color; ii) *E. coli* form circular, smooth and glossy colonies; iii) A lawn of *E. coli* forms a solid, smooth, glossy mass of cells.

Over the incubation, a single bacterium will grow and divide many times to form a colony. On your streaked plate, you should see something similar to what you saw in Virtual Bioengineer Canvas Kit Edition. There will be areas of high concentration of bacteria and areas where you can see single dots. These dots are called bacteria colonies!

In this exercise, all of the bacteria on your LB agar plate should be the same color or colors as written on the tubes of cells you received in the kit. *E. coli* have a smooth, glossy morphology and an overall circular shape (Figure 3-9).

If you see growth on your LB agar plates that is:

- A different color than you expect, this is likely contamination from your environment.
- A fuzzy growth is likely mold that was in the air
- An irregular shaped colony (not a circle like *E. coli* colonies are) this could be skin bacteria that you transferred into your experiment.

If you see any contamination in your plates, you should put them in your inactivation bag immediately so they can be inactivated at the end of your experiment (Step 13).

Once you see your results, you can move onto using these colored bacteria as your paint for the bioart you will make on your remaining petri dishes.

When creating your paintings, you will also be practicing a new skill called re-streaking bacteria. Re-streaking is used to grow more of an experiment's results for further experimental needs. In your case, this means that you will use the grown bacteria on your incubated plates as the 'source' bacteria for streaking and creating bioart on your remaining plates, instead of using the cells in the tubes that came with your Canvas Kit. Note that typically when you use re-streaking in a laboratory experiment, you only touch your inoculating loop to a single colony, but for painting, you can touch many colonies at once and/or the lawn of bacteria.

Note that if nothing grew on your petri dish, if the colonies did not change colors as expected or if your experiment was contaminated, discard the plates in the inactivation bag and repeat Step 7 & 8 on a new petri dish using the tubes of bacteria paint (cells) as your source bacteria. Remember to make sure to dip the yellow loop into the tube of bacteria paint in order to transfer the bacteria to your LB agar plate. You should be able to see some trace of your loop's trajectory on the surface of the agar if you hold it in the light. It will look like a "wet" trace. The individual bacteria themselves are too small to see and will require time to grow into colonies for you to verify success.

Growing *E. coli* *Going Deeper* **3-3**

Humidity: Bacteria need water/humidity to operate. The cell machinery and chemical reactions that occur in *E. coli* require the presence of water. All of the food, minerals, and salts in the cells are dissolved in water, enabling them to move and interact with other cellular molecules such as lipids, sugars, proteins, and DNA.

If the cells dry up, then cells enter "suspended animation" or break apart and die. In times of stress, such as dehydration, some bacterial species other than *E. coli* enter a long-term survival mode where they sporulate, becoming a spore that can remain in suspended animation for thousands if not millions of years.

Flipping plates so the bottom is up is a conventional technique for preventing LB agar plates from drying out. In the warm incubator, water in the LB agar evaporates. When a plate is inverted, that evaporating water does not escape as easily. It also keeps the surface of the LB agar moist. This aids in bacterial growth.

Temperature: Thermal energy (temperature) has a significant influence on the activities happening within cells. A cell's temperature can influence how fluid the cell membranes and cellular components are, speeding up or slowing down chemical reactions. A change in temperature can have profound effects on how fast an organism grows, along with the processes that go on within a cell. As you will see in later chapters, temperature can play a role in whether parts of the DNA blueprint (*e.g.* genes) are read by the cell. You can try growing your cells at 30 °C instead of 37 °C to see the difference in how fast the cells grow!

Why do you grow *E. coli* at 37 °C? *E. coli* bacteria have evolved inside of the large intestines of mammals like humans. Because our body temperature is 37 °C, the microorganisms have evolved to grow most efficiently at this temperature. In optimal conditions, *E. coli* will divide every 20-30 minutes. That's about twice every hour. After 12 hours (about 24 divisions) a single bacterium could divide into 16,777,216 bacteria. However, because of space and food constraints, as bacteria grow more densely, the pace of division slows.

Step 10. Painting living art (bioart) with *E. coli* bacteria

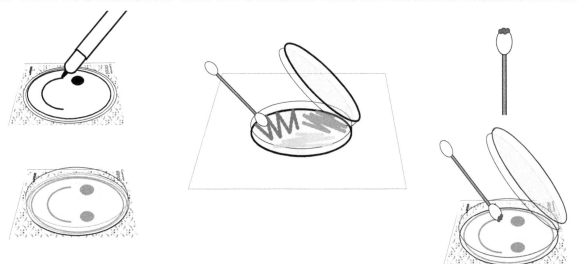

Figure 3-10. Bioart can be made by painting colorful bacteria onto an LB agar petri dish. Use the bacteria paint you streaked and incubated as your 'paint' for your art.

After incubation, verify that the bacteria paint has changed color before starting this step. If it is not yet colorful, incubate longer or refer to the manufacturer's instruction manual. The colors develop best at 37°C. If you are using a DIY incubator where the temperature is not 37°C or very close to it, your colors may not develop fully - they will stay in shades of pastel. You can use them as pastels paint.

Take two LB agar Petri dishes (from Step 1) out of the refrigerator. For each petri dish:

1. Draw your own design on one of the blank stencils. Place one of your unused LB agar plate on top of your stencil or the pre-made image stencil. Simple images without many details work best for the first time!

2. Take your petri dish streaked with colorful bacteria from your incubator. This is your painting palette. Using your 'bacteria paintbrushes', dip one into the paint on your painting palette. Paint your art by carefully streaking across the surface of the LB agar of your new plate, following your image stencil. When doing bioart, it is okay to dip your painting instrument into the bacteria more than once, but since you will be making more painting s on remaining petri dish(es) keep some paint and paintbrushes for later.

Remember to verify that your bacteria paintbrush touched the paint. After you have painted your image on the agar, you should be able to see some evidence of your work as a 'wet' trace if you hold the petri dish in the light. The individual bacteria themselves are too small to see and require time to grow for you to verify success. Painting with bacteria is hard because it is like painting with invisible ink that appears days later!

3. You can choose to paint another agar plate right away or save it for later. After you are done painting for the day, put used items like loops and cotton swabs in the inactivation bag. Close your painting palette petri dish and place it in a resealable bag (You can use the same one that has the unused agar petri dishes). Place it, along with the remaining unused petri dish, and bacteria paintbrushes in your lab refrigerator or in a sealed container that goes in the refrigerator (Chapter 2). You will incubate your painted plate(s) in the next step.

Step 11.   Incubating *E. coli* cells

Once again, you will be incubating petri dishes. This time, you will incubate your painted petri dish. Turn on the Incubator of your Minilab to 37°C. As the incubator heats up, you should flip your LB agar plates so that the bottom is up and the lid is down. Following the manufacturer's instructions, place the flipped LB agar plates with *E. coli* bacteria onto the paddle, cover with the humidity chamber, then place into the incubator. Turn on the timer to better monitor growth.

Incubate the petri dishes for ~24-48 hours. You will start to see growth after ~12 hours, and the bacteria will start to change color shortly after.

Step 12.   Viewing & Preserving your bioart with a Keep-it Kit

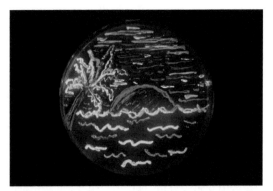 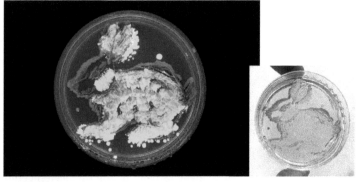

Figure 3-11. Left: Bioart made with fluorescent bacteria by Nathan Shaner, photo by Paul Steinbach. Created in Nobel laureate Roger Tsien's lab. Right: Bunny bioart made with Amino Labs' Canvas kit under black light and natural light.

After 24 to 48 hours of incubation, have a look at your work. You've created bioart! Since you have one LB agar plate remaining, you can repeat the painting exercise on this plate using what you learned after seeing your bioart results. Bioart is not an easy art form to master!

If you have a Keep-it Kit, you can now download the instructions and complete the Keep-it Kit exercise. As part of this exercise, you will let your plates dry so that there is very little moisture on the surface of the petri dish. This will be followed by embedding your cells in a clear resin and allowing it to harden over 24-48 hours.

Step 13.   Clean-up and inactivation

Figure 3-12. Use the inactivation bag to safely dispose of your experiment materials.

As you were going through the exercise, you placed used materials into the **inactivation bag.** When cleaning up at the end of this experiment, you should place all of your materials that have come into contact with bacteria into the inactivation bag, including your plates (don't forget to photograph them before destroying/inactivating them!). Note that the paper packaging for the inoculation loops can be disposed of in the regular garbage. You can also put your used gloves into the inactivation bag if you believe they've come into contact with the bacteria. All of your tubes should be placed in the inactivation bag with the lids off so the bleach water can enter them.

Follow the steps provided by the manufacturer:

1. If you haven't set up your discard container (Chapter 2), place your inactivation bag into a bowl, bucket, or similar type container to hold it upright and to contain any liquid if you accidentally puncture the bag.

2. Use your 'sterile water' bottle after Step 6. Add one full bottle **(250 mL) of concentrated bleach** to the bag. Be careful not to spill any bleach on yourself as it is caustic. Wear safety glasses & gloves!

3. Add 4 to 6 "sterile water" bottles of **warm water (total of ~1 L)** to the bag. The goal here is to submerge all the materials so that they are all in contact with the bleach water solution.

4. Seal the bag tightly and gently massage or swirl the items to ensure they are in contact with the bleach.

5. Let it stand for 24 hours or longer. This will inactivate the samples. In other words, it will kill the bacteria and break down DNA.

6. Afterward, empty the liquid in a toilet by clipping a corner of the inactivation bag with scissors (while over the toilet) and place the solid waste in the trash can. Cut only a small hole in the corner of the bag so that the solid materials do not escape.

CONGRATULATIONS!

on completing your second experiment!

Congratulations on growing *E. coli* cells on LB agar plates that you made yourself. This is the first major step toward becoming a Genetic Engineering Hero.

The hands-on exercise was packed with new information and essential skills. In the following section, you will take a deeper look at the fundamentals of *E. coli*. We will...

- Look at the history of the K12 *E. coli* cells you grew.

- Examine the structure of *E. coli*: what they look like, what they're made of, what helps them grow.

- Just as you learned about the DNA macromolecule in Chapter 1 *Fundamentals*, we will look at three other important macromolecules: sugars, proteins, and lipids.

This section features a tour of a K12 *E. coli* 'microfactory' which in many ways is like a traditional factory that you might see in your everyday life.

Fundamentals: *E. coli* Cells

Introduction to "Lab" *E. coli*

Escherichia coli (*E. coli*) is generally associated with hamburger meat and infection. However, there are many different strains (types) of *E. coli*, and most are not pathogenic (Figure 3-13). In fact, *E. coli* aids in digestion and helps to produce amino acids and vitamins for our bodies. Although we don't have the evidence yet, it is possible that every human has at least some *E. coli* bacteria as part of their large intestine microflora.

E. coli is the most widely used organism in scientific research, and is one of the simplest organisms that can be used in genetic engineering. If you are familiar with using software, you can think of *E. coli* as a "Lite" version of a software suite. Compared to other organisms like yeast cells or mammal cells, which have many more cellular "bells and whistles", *E. coli* is simple to engineer, grows fast (divides every 30 minutes), and is reliable (Figure 3-14).

E. coli was first observed in 1885 by the German-Austrian pediatrician Theodor Escherich, after whom the bacteria is named. *E. coli* may have a prevalence of around 0.1% of "normal gut flora". There are more than 700 strains of *E. coli*, including the lab strain you will use throughout this book. Only a few versions of *E. coli*, such as strain O157:H7, are pathogenic due to the production of bacterial toxins. These pathogenic strains of *E. coli* are Risk Group 2 (RG-2) organisms and should only be explored using Containment Level 2 Laboratory guidelines and conditions, which requires government approval.

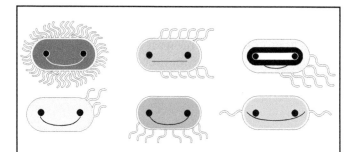

Figure 3-13. There are many different 'versions' of *E. coli*. Scientists are responsible for making some to better understand how *E. coli* functions, but many have also naturally evolved to be different.

On the opposite end of the spectrum, the *E. coli* strain DSM 6601 (also known as Nissile 1917), was isolated in 1917 from a World War I soldier who had a resistance to diarrhea, and who was known for having a "strong stomach". *E. coli* DSM 6601 has now been used as a probiotic for over 100 years!

The strain you will be using throughout this book is called the "K12" strain of *E. coli* - this *E. coli* is probably the most studied and well-understood organism... ever. Originally isolated at Stanford University in the

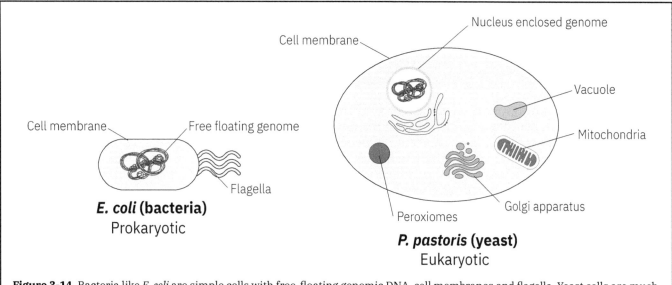

Figure 3-14. Bacteria like *E. coli* are simple cells with free-floating genomic DNA, cell membranes and flagella. Yeast cells are much more complex and have other specialized cell structure/compartments that the cells use to produce energy, make or break down fats, a nucleus to store DNA and a golgi apparatus to help fold and package products and machinery that the cells produce.

1920s, the K12 strain has gone through much evolution and manipulation in labs to become the lab strain we take for granted today.

Recall learning about plasmids at the end of Chapter 1; the short circular DNA helix loops that are akin to a USB stick for cells. In addition to a 4,600,000 nucleotide "main" chromosome, natural *E. coli* have a large plasmid called an "F plasmid" (short for "Fertility factor plasmid"). This plasmid enables one bacteria to share DNA with another through a process called bacterial conjugation - it allows the bacteria to communicate with one another. In lab strains of K12 *E. coli*, these F plasmids have been removed by scientists, making it very difficult for the bacteria to share genetic information. This substantially reduces the risk of K12 bacteria sharing information with other organisms if they are released into the environment. It also prevents communication between organisms in your samples during your experiments. K12 *E. coli* is not only a scientifically-sound organism choice, it is also a responsible one.

As you will learn in coming chapters, genetic engineers sometimes use antibiotic resistance genes during the genetic engineering process. A question that sometimes arises in public discussion is whether the antibiotic resistance genes used in genetic engineering cause the emergence of "superbugs" resistant to all antibiotics. The short answer is no, and one of the main reasons for this is that lab strains of *E. coli* have been "gagged" by the engineers - they cannot really share their DNA with other cells thanks to the removal of the "F plasmid".

A second significant difference between Lab K12 *E. coli* and other forms of *E. coli* is that they no longer have a "lambda phage" infection. Lambda phage is a virus

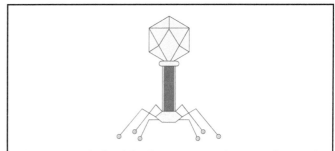

Figure 3-15. The lambda phage is a virus that can infect *E. coli*.

that infects *E. coli* and becomes part of its genome (Figure 3-15). Yes, even bacteria can get viruses and catch a cold! The original K12 strain has the lambda phage in its genome. Under the right conditions, the virus can become active, kill the cells and spread to others. Lab strains of K12 *E. coli* no longer have this infection, and you do not have to worry about your bacteria catching a cold during your experiments.

Throughout the 1950s-1980s, experiments were completed to first remove lambda phage and then the F plasmid, as well as a few other genes to achieve a version of K12 *E. coli* called "MG1655". Further tweaking to this bacteria has led to the most widely used *E. coli* K12 bacteria strains, called DH5α and DH10β. Labs have made hundreds of other versions of *E. coli* to study more about how cells work, but the DH5α and DH10β versions remain the most commonly used in genetic engineering. These are the *E. coli* strains you will be using and encounter in this book. For now, the K12 *E. coli* that you have already used in the hands-on exercise and whose ancestor was originally isolated from the feces of someone at Stanford University in California, about 100 years ago will be the model organisms we study.

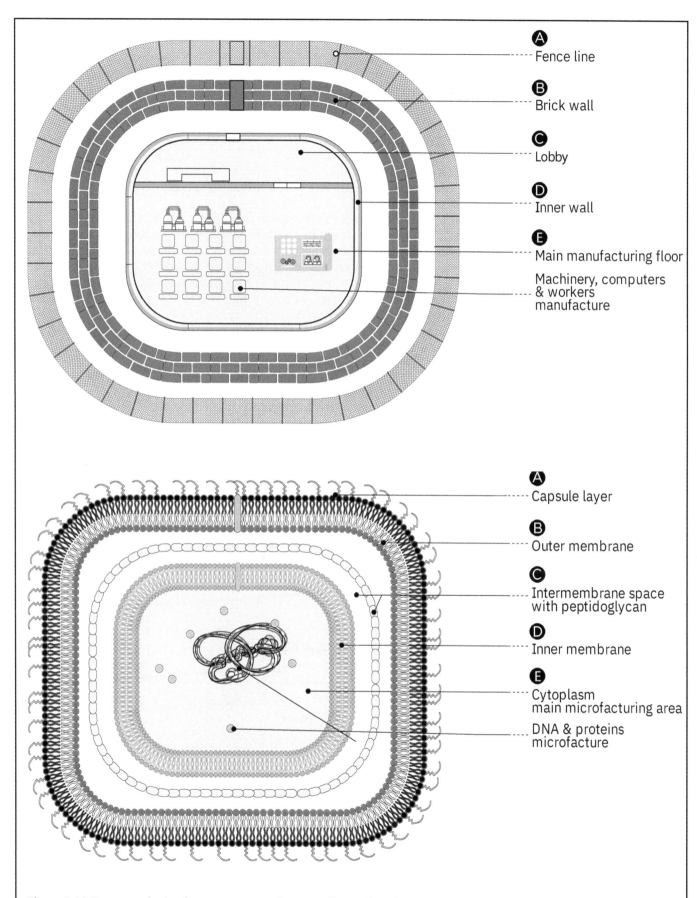

Figure 3-16. Factory and microfactory tour map. The tour will move from fence line to manufacturing floor; capsule layer to cytoplasm.

Insulin is a protein made in the pancreas and exists within the blood to regulate the amount of sugar that cells can absorb. When a person cannot make insulin in sufficient quantities, they have type I diabetes. Treatment of type I diabetes began in the 1920s after the discovery of insulin. Back then, insulin was extracted from barnyard animals by harvesting and grinding up their pancreases. Insulin was then extracted, purified and injected into humans with diabetes. While this form of insulin did treat diabetes, it had undesirable side effects, including allergic reactions, and required a lot of animal pancreases.

Visit https://amino.bio/insulin to be redirected to a website with a history of insulin. Here, you will find an image of a massive pile of pig pancreas that were used to produce only 8 ounces of insulin.

In the 1970s, the fathers of modern biotechnology (scientists Stanley N. Cohen and Herbert Boyer) developed methods to 'cut and paste' DNA from one organism to another. In 1976, Boyer founded a company called Genentech. Their first project was finding the DNA blueprint for human insulin so it could be inserted into *E. coli* bacteria. They were successful and created the first modern biotechnology product: human insulin. In 1982, humanity ushered in a new era of manufacturing - using bacterial cells as microfactories to deliberately produce important end products for a commercial purpose. It is estimated that over 100 million people rely on microorganism-produced insulin every year to stay healthy and alive. Now, hundreds of other medicines are produced using genetic engineering and biomanufacturing methods, both of which you will become familiar with in later chapters.

A Tour of the *E. coli* Microfactory

Now that you're familiar with the history and origin of the K12 *E. coli* that you streaked and painted on your LB agar plates, the rest of the *Fundamentals* involve digging deeper into what makes *E. coli* tick. Let's have a look "under the hood"!

In Chapter 1, we learned what makes up the macromolecule DNA: CHOPN. We saw that CHOPNS (carbon, hydrogen, oxygen, phosphorous, nitrogen, sulfur) have been classified as organic elements because they are commonly found in organisms. You can refer to the periodic table at the end of the book to refresh your memory. In this chapter, you'll see that all other macromolecules like sugars, proteins, and lipids also include CHOPNS in their make-up.

To make this information more relatable, we are going to use a cells-as-factory analogy, starting with a tour of the *E. coli* 'microfactory '. We call it a microfactory because when *E. coli* is used in genetic engineering, the cells literally become a factory. They read instructions (the DNA you insert into the bacteria), take in raw materials (the nutrients from the LB Agar), and create a product (like insulin from the *Industry Breakout: Insulin Going Deeper 3-5*).

The goal of this chapter is to understand the overall structure of an *E. coli* cell and the atoms, molecules, and macromolecules that it is made of. We'll look at five key parts of the microfactories (Figure 3-16):

- **The capsule layer (A):** the outermost protective shield of *E. coli*.
- **The outer membrane (B):** the primary exterior structural barrier.
- **The intermembrane space (C):** a key passage for entering and exiting the cell.
- **The inner membrane (D):** a structural barrier within the cell.
- **The cytoplasm (E):** the primary space in which cellular activities occur.

The Fence (A)

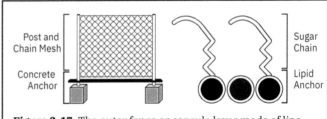

Figure 3-17. The outer fence or capsule layer made of lipopolysaccharides (LPS) protect the factories and are anchored in place.

The outermost property line of a factory is defined by a fence. This forms the outermost barrier of the factory, protecting the entire property from outside interference. The fence posts are anchored into the soil using concrete (Figure 3-17).

In the microfactory, we also see a "fence line" at the outer edge of our K12 *E. coli* cell. *E. coli* cells have an outer perimeter called the "capsule-layer" or "slime layer." It has a similar function to a fence: to separate the body of the cell from the outside environment. The capsule layer is indeed slimy and protects the cell by preventing chemicals (like antibiotics) and other organisms from entering the cell or disrupting the cell membrane. The capsule layer also determines how well a cell can keep water inside.

Similar to how a factory's fence line is anchored to the ground, the capsule layer of the *E. coli* is made of strings of sugars that are anchored to the underlying cell membrane by lipid macromolecules (Figure 3-17). The hybrid molecules that make up the capsule layer are made of both a sugar and a lipid. This molecule is called lipopolysaccharide (LPS). Look back at figure 3-16 to see how the LPS sugar is anchored in the outer membrane. We will talk more about lipids when we visit the cell membrane. For now, we will investigate what brings the "slimy" to the slime layer - sugars.

Sugars are part of a class of macromolecules called carbohydrates. *Carbo* meaning they have lots of carbons (C) and *hydrates* meaning they have lots of hydroxyls (OH). These molecules are important. They make up *E. coli*'s cell structure, and they are the 'fuel' that drive *E. coli* cell metabolism. Carbohydrates are typically made up of CHO, with carbon being the key atom that links to other carbon, hydrogen and oxygen atoms.

Sugars can be made from a variety of carbons, hydrogens, and oxygens. One of the most commonly known sugars, glucose, has six carbons that make up its

"backbone" (Figure 3-18 (left)) and, when dissolved in water, a glucose molecule folds up to form a ring structure (Figure 3-18 (right)). In Figure 3-18 (center), you will find a glucose molecule that does not have its carbons and hydrogens shown. See the *Reading Molecule Drawings Pro-tip* below to learn more.

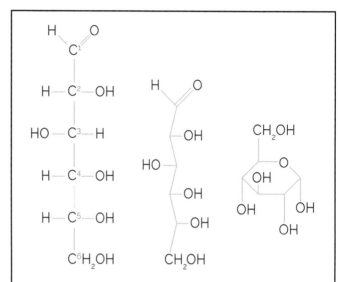

Figure 3-18. Glucose is a simple sugar with six carbons that are also connected to hydroxyl (OH) groups. On the left is a linear depiction of glucose with all atoms visible. When glucose is in liquid, like within a cell, it circularizes into the ring structure on the right.

The hybrid molecule lipopolysaccharides (LPS) gets its name from 'lipo, meaning lipid (fat), 'poly', meaning many, and 'saccharide' meaning sugar. Each LPS is a string of sugar rings permanently connected to a lipid anchor (Figure 3-16). In the cell's capsule layer, it is the strings of sugar rings that slip and slide on one another, making the surface of the bacteria 'slimy'.

Reading Molecule Drawings *Pro-tip*

As a way to make drawing chemical molecules easier, scientists decided that, since carbon is a highly used element in biochemistry, it could be left out of chemical drawings. When you see a molecule drawing like Figure 3-18 (right), you must know that any vertices without any letters in the ring are carbons. For example, the glucose in Figure 3-18 (left) has all of the elements drawn and labeled, and some, such as the C1-C6 down the backbone don't have to be labeled, as shown in Figure 3-18 (center). This may seem a little complicated, but as you learn and practice chemistry, it will become easier.

Scientists took this further. Since they know the rules of what can bind to carbons (each carbon binds to four other atoms in most circumstances, or requires four bonds), hydrogens are not annotated either. If you look at Figure 3-18 (center and right), many hydrogens (H) are also missing. While they are actually present in the molecule, they have been deliberately removed from the drawing for "simplification". Every carbon "vertex" with an OH connected also has an H connected in the opposite direction (purposefully not shown).

You may recall the reasons why lab strains of K12 *E. coli* are safer to use: i) they cannot effectively share genetic information and; ii) they do not have a phage infection. Here is another reason: K12 *E. coli* are missing some sugars in their LPS called the 'O antigen'. The O antigen is usually part of the sugar ring string on the LPS, serving as a first line of defense for the cell against the environment. Not having the O antigen means that the first defense layer of the microfactories is severely disrupted and makes the *E. coli* highly susceptible to dying from antibiotics, chemicals, acids, surfactants, and dehydration. It is as though the factory fence is only 1 foot tall with very large mesh holes. It offers some protection but not great protection. For this reason, K12 *E. coli* are not good at surviving outside the petri dish and are great to use in experiments.

The Outer Wall (B)

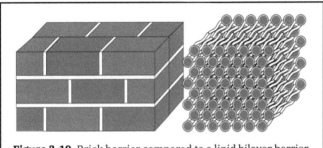

Figure 3-19. Brick barrier compared to a lipid bilayer barrier.

Have a look at your tour map (Figure 3-16). We have passed through the fence/capsule layer, and now we're moving on to the next stop - the outer factory wall.

The outer factory wall has one primary function - to be a strong barrier between the inside of the factory and the outside environment. The factory wall should protect the interior from rain, snow, hail, and even a tornado. The walls of the factory are made of solid red brick that is reinforced with iron.

The outer factory wall in our *E. coli* is the outer cell membrane (Figure 3-19). The function of the cell membrane is not only to protect the cell from the outside environment, but it also acts as a container for all activity inside the cell. Without the cell membrane, all of the cell insides would spill out. There would be no cell!

As a general rule, the membranes of *E. coli* cells are made up of lipids (Figure 3-20). Lipids are essential for keeping cell structure. Lipids are made up of CHO, CHON, CHONS, or CHONP, and in general, have the

common characteristic of having a "tail group" and a "head group" (Figure 3-20), just like the surfactants you used for cell lysis in Chapter 1.

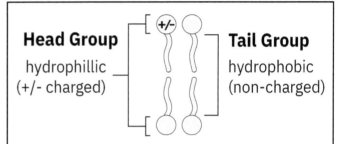

Head Group
hydrophillic
(+/- charged)

Tail Group
hydrophobic
(non-charged)

Figure 3-20. In membranes, the head groups are charged molecules that interact well with water and other charged molecules. The tail groups are non-charged, and they do not like to interact with charged groups but like interacting with other non-charged groups. For this reason, many biological membranes are made of lipids that form of a lipid bilayer.

The "head group" segment of the membrane lipid is hydrophilic (water-loving). This means that this is the part of the lipid that interacts with the water environment inside or outside of the cell. The head group is hydrophilic because it is charged, making it interact with water, which is also charged. The structure and composition of the "head group" can vary greatly. For example, in Figure 3-22, a negatively charged phosphate molecule is present. In other lipids, the phosphate is replaced by a sulfate or other chemical group.

The "tail group" segment of the membrane lipid is hydrophobic (water avoiding) (Figure 3-20). This means that this part of the lipid prefers not to make contact with water. It is hydrophobic because it is uncharged and does not like interacting with charged molecules or atoms. Because of this, it prefers to interact with the tail groups of other lipids (Figure 3-20 and 3-21). This difference in binding preferences leads to a structure called a lipid bilayer, where the water-loving head groups face outwards into the watery environment, and the water-avoiding tails face inwards toward other water-avoiding tails to form a protective barrier (Figure 3-20 and 3-21). When millions of lipids are produced by a cell, the layer creates a robust barrier that separates the inside of the cell from the outside environment (Figure 3-16).

During the first point of the tour, we talked about LPS – the chain of sugars that make up the slimy part of the capsule layer. However, those sugars are bound to a lipid anchor called "Lipid A". Lipid A has a hydrophobic tail that fits snuggly with the other "tail groups" inside the outer lipid bilayer at the surface of the cell. This hydrophobic tail is the anchor that holds the entire LPS molecule in the membrane.

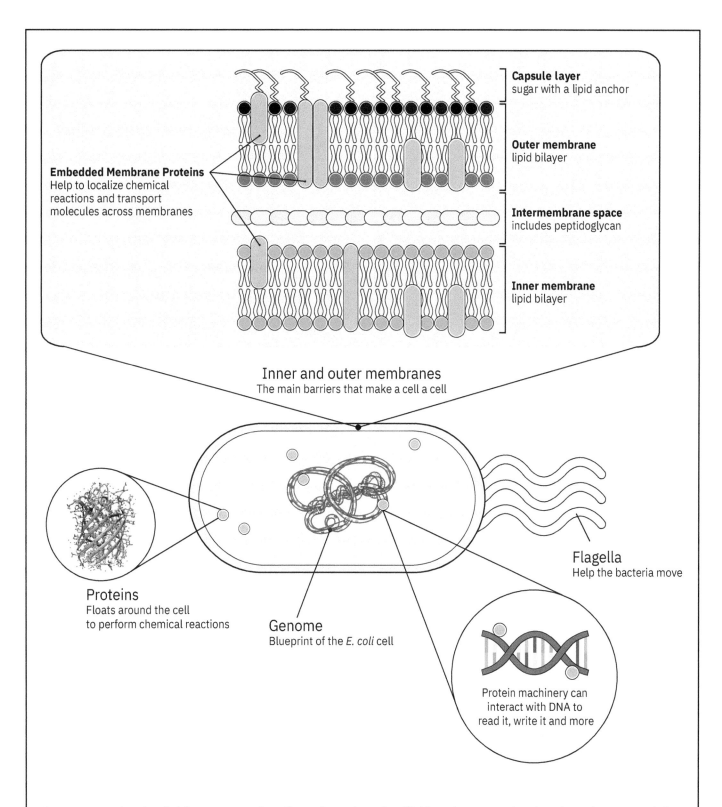

Capsule layer
sugar with a lipid anchor

Outer membrane
lipid bilayer

Embedded Membrane Proteins
Help to localize chemical
reactions and transport
molecules across membranes

Intermembrane space
includes peptidoglycan

Inner membrane
lipid bilayer

Inner and outer membranes
The main barriers that make a cell a cell

Proteins
Floats around the cell
to perform chemical reactions

Genome
Blueprint of the *E. coli* cell

Flagella
Help the bacteria move

Protein machinery can
interact with DNA to
read it, write it and more

Figure 3-21. Rather than brick or concrete, the cell membrane is made of lipids and sugars. Rather than doors that open and close, proteins are embedded in the membrane and act as "tunnels" that help specific molecules to enter and exit. The blueprints of the cell are not made of paper, but DNA. Small proteins made of amino acids are the cell machinery and scaffolding of the cells and readily interact with DNA. Flagella are a whip-like microstructure made of protein that help cells to move around.

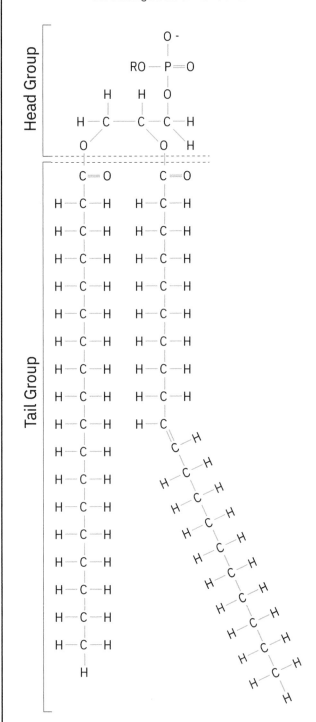

Anatomy of a Lipid
The Building Blocks of Membranes

Head Group

Tail Group

Figure 3-22. There are many kinds of lipids that make up the inner and outer membrane of *E. coli* cells. There is a general structure to these lipids, and that includes an uncharged hydrophobic tail and a charged head group. As in Table 1-1 in Chapter 1, the tail groups prefer to interact with each other, while the head groups prefer to interact with one another and the environment because environments on earth generally have lots of water. These chemical bonding preferences lead to lipid bilayers as well as micelles that you saw in Chapter 1.

The Lobby (C)

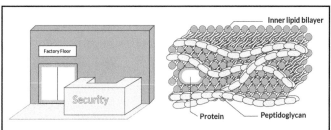

Figure 3-23. Factory lobby *vs.* the intermembrane space of a cell. At your back is the outer brick wall and outer membrane. You can see the inner wall/inner membrane space just beyond the lobby.

As we move into the factory, we pass through doors into the lobby (Figure 3-16). Cells also have doors, in a way. Some tubular-shaped proteins cross through the cell membranes and act as "portals", serving as the entry and exit points of the cell. This is where food enters the bacteria, and waste exits the cell. We will be learning in-depth about proteins very soon.

Once you enter the lobby (Figure 3-23) you have also entered the building. However, you are not yet on the factory floor. The lobby is a transitional location and acts as an extra checkpoint. In this lobby, there might be a security guard keeping an eye open for any trouble. The temperature, humidity, and air quality in the lobby is controlled but may be different than that of the factory floor. Nevertheless, it is much closer to those conditions than the outside environment.

The lobby-like space of K12 *E. coli* is reached by passing through the outer membrane. This 'lobby' is called the intermembrane space (Figure 3-16). and it exists between the outer cell membrane and the inner cell membrane. It is also a 'security checkpoint' since most molecules that enter or exit the cell must go through this space. In the inter-membrane space, there's an additional physical barrier called the peptidoglycan. The peptidoglycan is a mesh-like protective barrier between the inner and outer membranes of *E. coli* cells (Figure 3-21 and Figure 3-23). The chemical environment of intermembrane space (salt concentrations, water molecules, and other cellular molecules) is more similar to the inside of the cell than to the outer environment. In other words, the intermembrane space is a transitional space that helps to maintain a controlled environment inside the cell.

Peptidoglycan is made of sugars (N-acetylglucosamine) that are connected to short chains of amino acids (alanine, glutamine, lysine, glycine) to form the interconnected mesh. As you now know, cells can create hybrid macromolecules. These are a combination of different macromolecules, for example, a sugar bound to a lipid (to make LPS).

Peptidoglycan is a mix of sugars and amino acids, and you'll learn about amino acids soon. Over billions of years, cells have been doing 'molecular mash-ups' to find new and interesting molecules that help them survive. What is really cool, is that they are almost exclusively made up of CHOPNS!

The Inner Walls (D)

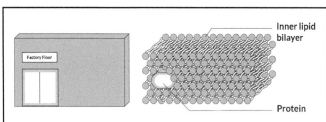

Figure 3-24. Factory inner wall *vs.* the inner membrane of a cell. You have passed through the lobby.

Have a look around this beautiful lobby - we can very clearly see the outer wall we just came through. We can also see a number of inner walls and passageways. These inner factory walls also play a very important role. Some parts of the factory may need to be warmer, colder or more humid than the lobby. These inner walls and passageways help to regulate the conditions in these different areas while contributing to the structural integrity of the building. (Figure 3-24).

The inner wall of an *E. coli* cell is the inner membrane, also made of a lipid bilayer (Figure 3-24). Just as you saw with the outer membrane, the inner membrane is comprised of a variety of lipids that have 'hydrophilic heads' and 'hydrophobic tails'. Just as the slime layer is anchored to the outer membrane, the inner membrane also serves to anchor proteins and sugars.

There are many different kinds of lipids that the cell makes in order to create and maintain its protective membranes: phospholipids, sphingolipids, bolalipids, and lipid A, to name a few. All of the lipids and proteins that make up the inner cell membrane contribute to its structural integrity and help to encapsulate and protect the next stop in our tour, the factory floor.

The Factory Floor (E)

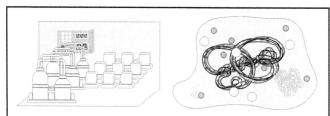

Figure 3-25. The factory floor and the cell cytoplasm contain the machinery and blueprints necessary to "run" the micro/factory.

Let's pass through the doors from the lobby and down the hall to the factory floor (Figure 3-16).

This is where it all happens! Look around, you'll see shiny machines humming while raw materials are processed and refined into usable products. 3D printers are reading factory blueprints to make a variety of valuable objects. Quality control machines scan items to ensure the size, weight, and composition are correct before releasing them. There are even small autonomous robots buzzing around, transporting items in the right place at the right time.

You probably noticed as soon as you entered - it is quite warm in here. The temperature has to be reliably maintained at 37°C. If you increase the temperature above 40°C or below 35°C, these beautiful machines start to shut down and make mistakes.

If you look closely, you'll notice that this factory is actually making and assembling the structural components of the factory itself, including the machinery! Wow, this is a factory of the future, a factory that creates new versions of itself! Amazing! That is what it is like to be in the cytoplasm of a K12 *E. coli* cell! All of the machinery you see are proteins. Let's learn about proteins, the cellular machinery that makes the miracle we call life possible.

Proteins are both the 'machinery' and 'scaffolding' (structure) of cells. Proteins help cause chemical reactions to happen through enzymatic reactions (Chapter 6). Some proteins read DNA (Chapter 4), while others copy and create DNA. Proteins are even essential for creating other proteins (Chapter 5). Proteins break molecules and other proteins apart, combine molecules and reshape molecules. For example, protein enzymes are why a lipid and a protein combined to form peptidoglycan in the intermembrane space. They also caused the sugar and lipid of LPS to become connected and bind into the outer membrane.

Proteins are also an essential part of the structure and function of cells. Like central beams that support the factory roof, strings of proteins can form the scaffolding of cells to help give cells structure and shape. As we saw during our tour, proteins can be the 'doors' and 'passageways' through the inner and outer membranes (Figure 3-24).

Proteins do a lot, but what are they and what are they made of? Just as DNA is a string of smaller "building block" molecules called nucleotides, proteins are a string of smaller molecules called amino acids (Figure 3-26).

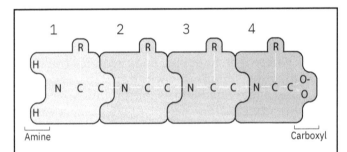

Figure 3-26. Four amino acids linked together to form a chain called a peptide. A long peptide of 20 amino acids or more is a protein. "R" represents the variable side-group of the amino acid that you can see more of in Figure 3-30.

While the nucleotides that make up DNA are made of CHOPN, amino acids are made up of CHONS, and amino acids string together to make proteins. Proteins do not form a double helix, that is a special characteristic of DNA. Instead, they form a single long string of amino acids that can fold into a three-dimensional shape like when your headphone wire tangles up on itself!

Comparing again to DNA, at the molecular level, the string of nucleotides is joined together because of a sugar-phosphate backbone (Figure 1-17). Each nucleotide has its own phosphate and deoxyribose sugar, however, when combined with other nucleotides

to form a string, you see that the sugar-phosphates are attached to other sugar phosphates to create the backbone of DNA. The backbone of an amino acid's string follows the same principle, but the chemical groups are different. Amino acids are made of three different chemical groups - an amine (NH_2 group), a carboxyl (COO- group) and a unique group called a "side-group" (Figure 3-27).

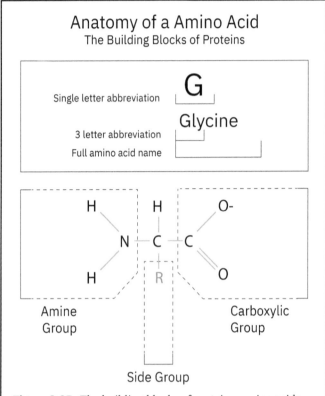

Figure 3-27. The building blocks of proteins, amino acids, have important characteristics. An amine group, a carboxylate group and a special group that makes them unique, the 'side-group'.

Scientists call a nitrogen that is bound to two hydrogens (NH_2) an "amine." When a carbon (C) is bound to an oxygen (O) and another oxygen with an hydrogen (OH), it is called a carboxyl group, short for carboxylic acid. As the amino acid string is created the amine-carboxyls are slightly altered so that at the very beginning of the string there is an amine, then there is a repeating nitrogen-carbon-carbon (N-C-C) backbone, and at the end is a carboxyl (Figure 3-26).

A short string of amino acids is called a peptide. When the chains include more than ~20 amino acids, they are called proteins. The chemical structures of all of the different common amino acids can be found in Figure 3-30, and these amino acids can be strung together in any order!

One thing you will notice is that when amino acids are strung together to form a peptide, the first amino acid gets to keep the amine group. An abbreviation for an amine is N for its nitrogen. This is why the start of the peptide is called the N-terminus (Figure 3-26). You'll also see that the last amino acid in a peptide keeps the carboxyl group, which has the abbreviation C because of the Carbon. This is why the end of the peptide is called the C-terminus (Figure 3-26).

Perhaps the most interesting part of the amino acid is the "side group". Think back to Chapter 1: Just as DNA is made of conserved sugar-phosphate backbone (conserved means it is constant and repeating in DNA), and has a variable "A, T, C, or G" nitrogenous base - amino acids have a conserved N-C-C backbone and have a variable "side group". The side group is the small cluster of atoms that give each amino acid a unique characteristic. Amino acid side groups can be "water-loving" or "water-avoiding". They can be positively charged, negatively charged or uncharged. These different chemical characteristics mean they can interact with each other and their environment in many interesting ways. See Figure 3-30 to view different amino acids and the side chains drawn in orange.

Using the simplified bonding rules you learned in Chapter 1 (Table 1-1), think about how the amino acid chain folds as it is being created by the cell. For example, positive and negative charged amino acids will be attracted to one another. Whereas DNA has complementary pairs of nucleotides that they must bind to (A-T, C-G), amino acids do not have "complementary amino acids" that they must bind to. Rather, general bonding rules are what cause the amino acid string to fold up (in-depth discussion about bonding can be found in Chapter 6 *Fundamentals*). Recall that opposite charges attract (+/-) while same charges repel (-/- or +/+).

In Figure 3-28, you'll see a simplified example of how different side chains can interact with one another, causing attraction or repulsion and ultimately raveling up into a tight three-dimensional structure. In Figure 3-28, side groups with a positive charge are indicated with a '+', those with a negative charge have a '-', and side groups with no charge are marked by an 'O'. Following the simple rules, you'll see that these different charges drive the interactions. Remember, non-charged molecules like to interact with other non-charged molecules. In the real-world example, a negatively charged amino acid such as glutamate (Figure 3-30), can interact with a positively charged amino acid such as lysine (Figure 3-30).

You will learn a lot about the different functions of proteins throughout this book. If you're interested in seeing what a protein looks like, a computer-generated image can be found in Chapter 6 (Figure 6-17).

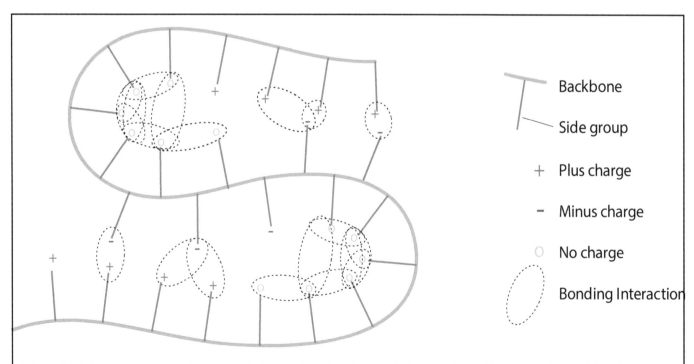

Figure 3-28. A string of connected amino acids folds up based on the chemical properties and bonding of the variable side-groups. Refer back to Table 1-1 to see some simple rules of bonding.

Fold-it! *Breakout Exercise*

How do you think this peptide chain would fold? Apply what you know about hydrophillic and hydrophobic bonds!

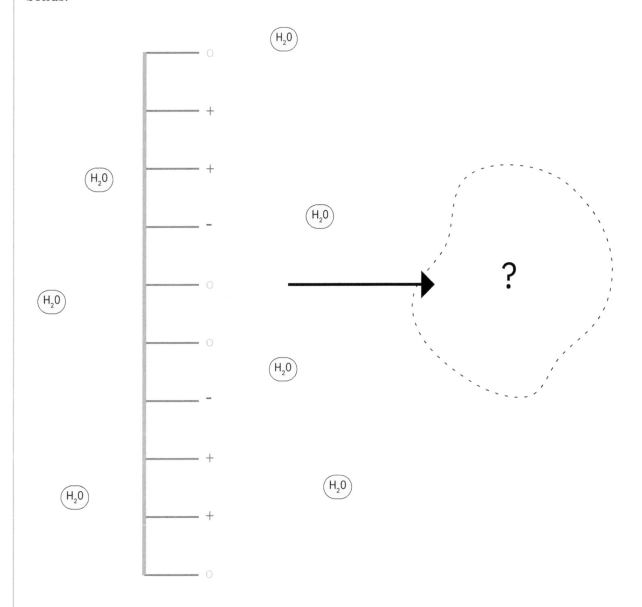

Remember, the solution to these breakout excercises can be found at <u>amino.bio/community</u>

Did you know? There is a game called *Fold-it* where you get to help the software fold proteins! This in turns helps scientists apply the resulting protein folding knowledge in their labs. Find it by searching "Fold-it science" online.

Inner Life of the Cell *Video Breakout*

Inner Life of the Cell was one of the first computer generated videos that explored the inner structure of cells. The creators of the video used information and data from research papers to guide its creation. Find it online and watch it. It is fascinating (short and long versions with or without narration exist). A key takeaway is that cells are very mechanical and factory-like.

Molecules in your everyday *Breakout Activity*

You have now started to learn about the molecules that make up the structures of *E. coli*. You may be interested to know that you and other organisms are made up of the similar stuff! Go to your cupboard and fridge to grab a box of cereal and milk. If you don't have any, use the internet instead. Look at the "Nutrition Facts" on the packaging. You will see these categories:

Fat - While the term "fat" includes many kinds of molecules, lipids are included in this category. Within your cereal and milk are lipids that are either from the membranes of formerly living organisms or were produced and secreted by living cells such as the mammary glands in cows.

Cholesterol - Cholesterol is a molecule made by cells that becomes embedded in the lipid bilayer of the cell's membrane. Cholesterol modulates how rigid or flexible the membrane is. We did not discuss this, but *E. coli* do have cholesterol in their membranes!

Carbohydrate - Carbohydrates make up most of your cereal. These include many different kinds of the sugars, like the glucose you learned about in Figure 3-18.

Proteins – Many different types of proteins that were involved in helping the organism grow and survive are left over after the organism has been harvested. These will be broken down into amino acids and reused by your body to make proteins.

Protein Catalysts in Chemical Reactions *Going Deeper* 3-7

What do you call something that causes a chemical reaction that otherwise wouldn't occur? A catalyst. When you consider the thermodynamics (energy changes) of a chemical reaction, you have to consider the reactants and the energy of the system. This decides whether a reaction will actually take place.

Imagine you're on a roller coaster and you are a "reactant" - something that will transform into a product given the right conditions. In this hypothetical roller coaster reaction, you start out as the reactant "anticipation and fear", which, given the right conditions, turns into "exhilaration and fun" (Figure 3-29).

As you're sitting in the roller coaster car waiting for the ride to start, you are very much "anticipation and fear". No matter how long you sit there, you will remain an "anticipation and fear" reactant. However, as the ride starts and you start to climb higher and higher, energy in the form of "height off of the ground" is added to the system. You, as "anticipation and fear" remain in that state and your stomach might begin to churn as the anticipation further builds. Eventually, with enough "height off the ground" energy, you reach the top of the roller coaster, and as the car lets loose, the roller coaster chemical reaction is in full swing! There is now enough energy in the system to propel you into the "exhilaration and fun" product state. The roller coaster falls straight down, your adrenaline explodes, and you SCREAM. Screaming is a by-product of the roller coaster reaction.

As your roller coaster car returns to the loading area, you have completely transformed into the "exhilaration and fun" state. All of the anticipation and fear have transformed into feelings of exhilaration and fun. You also experience some relief, another by-product of the roller coaster reaction. The key parts of the reaction were the reactants (you as "anticipation and fear"), the energy needed to start the chemical reaction called the activation energy (height of the roller coaster), and the products (exhilaration and fun, with some scream and relief). In the case of most chemical reactions that happen in cells, there is not enough activation energy, usually in the form of heat, in the cells to cause the chemical reaction to occur on its own. This is where protein enzymes become essential.

Protein enzymes are 'magical' macromolecules that can 'catalyze' the chemical reaction. This means that even though there isn't enough energy, they can still cause the reaction to happen - they have a nifty "hack" to make the reaction happen. In technical terms, the "hack" is that the enzyme catalyst lowers the 'activation energy' required to cause a chemical reaction. This can catalyze a chemical reaction that wouldn't normally happen. In Figure 3-29 this is represented as a tunnel that bypasses the tower.

Each enzyme has a specific purpose and can catalyze a specific chemical reaction. This is what makes biology so unique. Thousands of protein enzymes catalyze the chemical reactions necessary for life, including the creation of enzymes to catalyze more chemical reactions. Chapter 6 covers this in-depth.

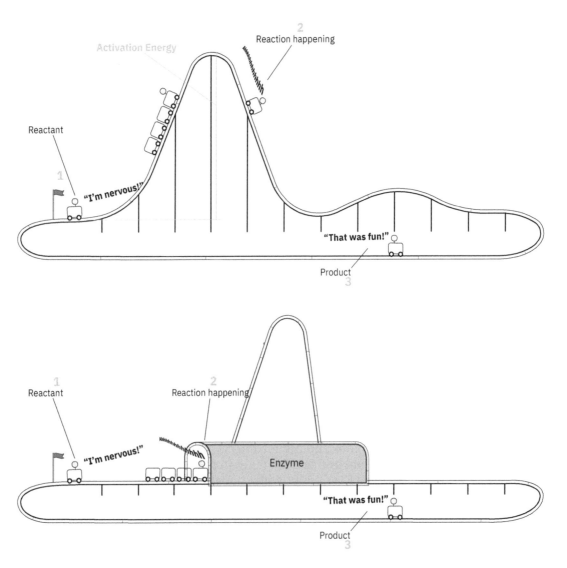

Figure 3-29. A normal chemical reaction requires reactants and activation energy needed to cause reactions. Enzymes lower the amount of activation energy needed and cause reactions to happen in cells that wouldn't normally happen.

What other microorganisms are used in genetic engineering? *Pro-tip*

Even though we will be using *E. coli* as our microfactories throughout this book, other organisms are also used as microfactories.

Living organisms that are genetically engineered to act as factories are mostly limited to single-celled microorganisms because they are easier to engineer, grow, and control genetically. Single-celled organisms grow and reproduce as singular entities, in contrast with multi-celled organisms like humans, where many cells grow together to form tissues, organs, and entire complex organisms.

Common single-celled organisms (or microfactories as we are calling them) used for biomanufacturing are listed in Table 3-2.

Considering the different classes of organisms, single-celled bacteria are generally the simplest to engineer as they have the least complex architecture and internal machinery. As you'll see in Table 3-2, bacteria are smaller and have much less DNA than yeast or mammal cells. This is another reason for starting out with *E. coli*, they are simply easier to program.

Table 3-2. Microorganisms Used for Genetic Engineering and Biomanufacturing			
Class	**Physical Size**	**Genome Size (MB*)**	**Examples**
Bacteria	1-3 um in diameter	0.6-5 MB in a single chromosome (~4,000 genes)	Escherichia coli (gram -), Bacillus subtilis (gram +), Synechocystis sp.(cyanobacteria)
Yeast	3-10 um in diameter	12 MB (~6,000 genes)	Pichia pastoris, Saccharomyces cerevisiae
Algae	10 um in diameter	~150 MB in 17 chromosomes (~14,000 genes)	Chlamydomonas reinhardtii
Mammals	15 um in diameter	~5,200 MB in 22 chromosomes (~24,000 genes)	Chinese Hamster Ovary cells (CHO)

MB: megabases = 1,000,000 base pairs (or nucleotides in length)

Mammals *vs.* Microorganisms:

The last row in Table 3-2 includes mammalian cells. But aren't mammals complex organisms? Aren't they made of many cells that form tissues?

The answer is yes! However, in the case of genetic engineering and biomanufacturing, many 'cancerous' cells (cells that can grow and duplicate without special signals from other cells) are used. Normally, cells can only grow and divide with the right signals from other cells. There are many check and balance systems in place to ensure that mammalian cells are unable to grow by themselves and can only complete the tasks they are specialized to do. But some cells can become cancerous.

In many cases, this is because their DNA was not correctly copied, allowing them to grow at their own pace, uncontrollable by other cells and by their environment. Genetic engineers are now turning some of these harmful and deadly cancerous cells into factories, often to create medicines that are saving patients' lives - even patients with cancer. TAKE THAT CANCER! Be aware that most human cell lines used for genetic engineering are considered RG-2 organisms and require special lab spaces and authorizations.

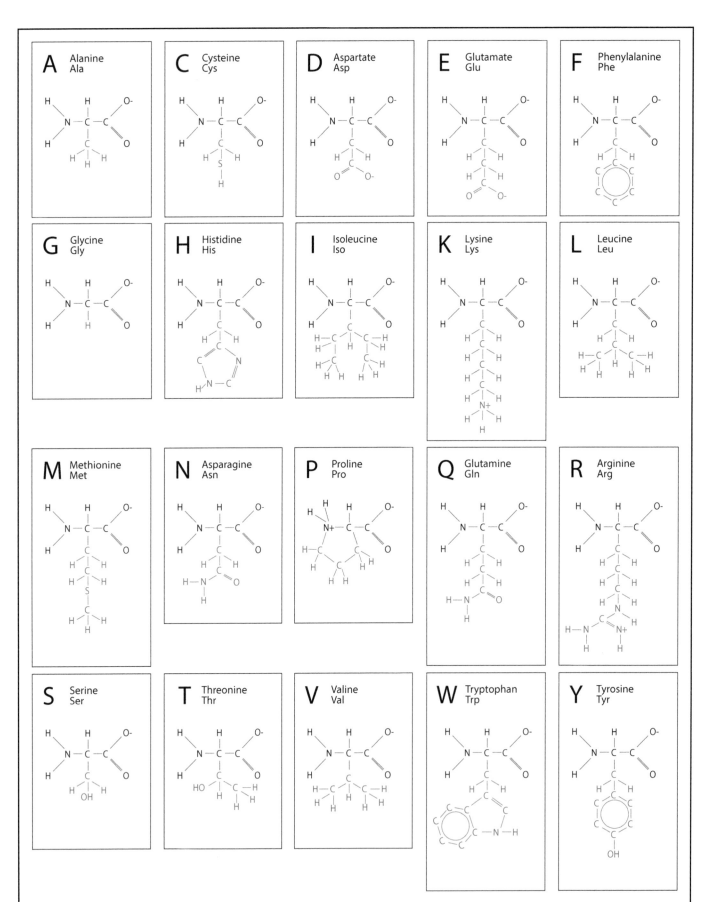

Figure 3-30. *E. coli* makes 20 amino acids that are then used as building blocks to create proteins chains. The orange region is called a "side group", and the green region is the "amine-carboxyl backbone".

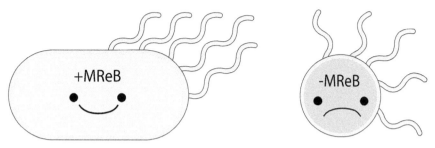

Figure 3-31. What causes bacteria to be spherical or rod shaped?

Proteins can determine cell structure. Why are *E. coli* bacteria rod-shaped and not spherical like other bacteria? It was thought that peptidoglycan and the outer cell membrane are responsible for the bacterium's rod shape. However, it has now been shown that several proteins, including one called "MreB", regulate the shape of *E. coli*. Genetically engineered *E. coli* that lack MreB becomes spherical in shape! A follow-up question is, why does *E. coli* benefit from being rod-shaped? Why has it evolved to have this characteristic? The answer is still unknown!

Summary and What's Next?

In this chapter, you grew real lab-strain K12 *Escherichia coli (E. coli)* cells on LB agar petri dishes. Congratulations! Knowing how to grow the organism you plan to engineer is the first step towards learning the genetic engineering process. This fundamental skill is used in virtually all your genetic engineering experiments. You will quickly become an expert at making LB agar plates and streaking microorganisms.

In the *Fundamentals* section, we dug deeper into where K12 *E. coli* come from and how they differ from the *E. coli* bacteria that live in your large intestine and the dangerous ones that can make you sick. We compared the K12 *E. coli* to factories and looked into the physical make-up of their cells: the capsule layer, outer membrane, intermembrane space, peptidoglycan, inner membrane, and the cytoplasm. Knowing the architecture of the cells is very important in your journey to becoming a Genetic Engineering Hero.

In addition to the physical structure of *E. coli* cells, you learned about the macromolecules that form them. You used the knowledge of nucleic acids gained in Chapter 1 to become familiar with sugars (a.k.a

saccharides or carbohydrates), lipids, and proteins. Collectively, these macromolecules are made up of CHOPNS, a group of atoms that scientists call organic elements.

In the upcoming chapters, the four different types of macromolecules (nucleic acids, lipids, sugars, proteins) will repeatedly come up. Your understanding of them will continue to grow and become more sophisticated. In Chapter 4, you will do your first genetic engineering experiment where you will take the skills you learned in this chapter to the next level by genetically engineering K12 *E. coli* cells with DNA plasmids, making them your own microfactories!

In the *Fundamentals* of Chapter 4, you will step one layer deeper into *E. coli* cells. You will go beyond simply what they are made of and more into how they function. Specifically, you will discover how cells read DNA blueprints to create other products through a process called transcription. Transcription requires some knowledge of all the macromolecules, so hang onto what you learned in the *Fundamentals* from this chapter!

Review Questions

Hands-on Exercise

1. What is LB agar used for? What are the two important ingredients in LB agar?

2. How do you know you have created your LB agar plates correctly?

3. How can you confirm that you actually dipped your loop in cells before you start streaking?

4. What are the steps for inactivation?

5. What is a plate?

Fundamentals

1. Where do K12 *E. coli* come from?

2. List three key deficiencies that K12 *E. coli* have compared to natural *E. coli*.

3. What are the five key locations in an *E. coli* cell and how do they relate to locations in a factory?

4. Describe what CHOPNS is.

5. What three macromolecules were described in this chapter and what atoms are each made of?

Chapter 4

Genetic Engineering Your *E. coli* Cells

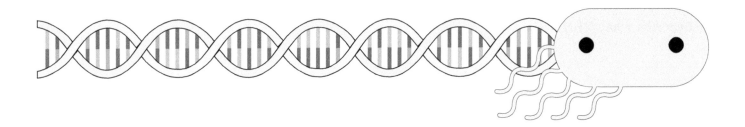

You have now extracted DNA from fruit cells. You have made LB agar plates. You have grown K12 *E. coli* cells using two different methods: streaking them onto a plate to get fresh bacteria paint and by painting them to create living BioArt.

Consider your BioArt for a moment. Did it turn out the way you expected? Painting with bacteria can be challenging because the cells can grow beyond where you've painted, distorting the image. As with all science and art skills, practice makes perfect!

In this chapter you are going to use the skills, you developed in Chapter 3 to take the next step on your genetic engineering journey. You are going to streak and grow cells in order to engineer them with DNA. You will change the actual blueprints of the cells so that they create something for you. This is the basis of genetic engineering.

In practical terms, you are going to learn how to insert DNA into cells and have the cells read and execute the instructions encoded in the new DNA. This won't require you to replace the genome of the K12 bacteria. Instead, you will be adding some additional DNA, called a DNA plasmid, into the cells. Once inserted, the bacteria will start to read and execute the DNA from the plasmid, just as they will continue to read and execute their normal genome "programming".

The *Fundamentals* in this chapter will build on everything you've learned in prior chapters. We will look at how a cell reads DNA and begins 'executing' the instructions to create something, a product, for you. The process of reading a DNA plasmid is called transcription, and it is one of the most important processes in biology. All of this happens on the "factory floor" in the cytoplasm of your K12 *E. coli* and mostly involves nucleic acids and proteins.

Just as you were introduced to the concept of genes in Chapter 1, in this chapter, you are going to learn first-hand how to insert genes into your K12 *E. coli* bacteria. Once the gene is inserted via a plasmid, we will learn how the cell reads the gene in order to start creating, or microfacturing something for you. Scientists also call this "expressing a trait". "Microfacturing" is a term that we use a lot throughout this book, but it is rather new to the field of genetic engineering! Since you imagine the cell as a microscopic factory (micro-factory), think of its activities as microfacturing.

Learning these subjects is the foundation of genetic engineering and critical to becoming a Genetic Engineering Hero. Once you understand how to insert DNA into a cell and how a cell reads and executes a "DNA program", you'll be on your way to learn advanced subjects like designing, making and executing your own "DNA programs."

Getting Started
Equipment and Materials

The **Amino Labs' Engineer-it Kit™** is contained in the ***Zero to Genetic Engineering Hero Kit Pack Ch. 1-4*** and includes all the required pre-measured ingredients. The Keep-it Kit™ is optional and both can be ordered separately at https://amino.bio/products.

Shopping List

Wetware kit: Amino Labs Engineer-it Kit™ (https://amino.bio)
Wetware kit: Amino Labs Keep-it Kit™ (https://amino.bio) (optional)
Minilab (DNA Playground)
Microwave
Laptop/Desktop PC for Virtual Bioengineer™ Engineer-it Kit Edition (https://amino.bio/pages/vbioengineer)

Instructions Overview

Day 1
1. Practice Engineering *E. coli* by completing the entire Virtual Bioengineer™ simulation
2. Make non-selective and selective LB agar plates
3. Streak *E. coli* bacteria on non-selective plates, and incubate in your DNA Playground for 12-24 hours

Day 2
4. Make chemically competent cells so your *E. coli* can take in the DNA plasmid
5. Heat shock and transform your DNA & *E. coli*
6. Recover your cells with enriched growth media

Day 3
7. Plate your transformed & recovered cells on selective LB agar plates
8. Incubate your plates for 24-48 hours

Day 4
9. View results

Day 5
10. Optional: Immortalize your first genetic engineering experiment (Keep-it Kit™)

Chapter Timeline Overview

Timeline to complete the hands-on exercise is:

Day 1: ~60 minutes followed by 12-24 hours incubation,
Day 2: ~60-90 minutes followed by 12-24 hours recovery
Day 3: ~20-30 minutes followed by 24-48 hours incubation
Day 4: ~15 minutes to view results. Optional: Prepare for preservation followed by 24 hours of drying time
Day 5: Optional: ~15 minutes to immortalize results followed by 24 hours of curing time.

Timeline to read *Fundamentals* is typically 3 hours.

Virtual Bioengineer™ Interactive Sim *Practice Breakout Session 1*

In Chapter 3, you completed Virtual Bioengineer™ Canvas Kit Edition and learned how to make LB agar plates, and practiced streaking/incubating K12 *E. coli* cells.

In this chapter, you will complete a different Virtual Bioengineer™ simulator, the Engineer-it Kit Edition, and also do those steps in real life. The simulation will familiarize you with the materials and procedure that you will be using to:

- Make selective and non-selective LB agar plates
- Streak cells and incubate them
- Collect colonies to make competent cells
- Add DNA and heatshock the cells to get your DNA into the cells
- Recover the cells
- Incubate the cells

Virtual Bioengineer™ is free and takes about 20-30 minutes to complete. It is recommended that you complete the simulation a few times. That way, when you do the hands-on exercises with real DNA and cells, you won't make any mistakes!

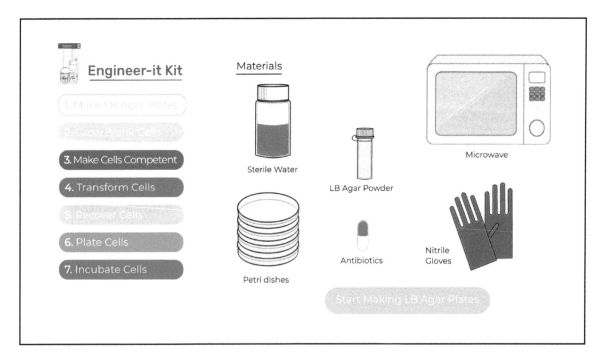

Learning Hands-On: Transforming K12 *E. coli* cells with a DNA Plasmid

Now that you have a good understanding of DNA, proteins, and the structure of *E. coli* cells, it's time to complete your first genetic engineering exercise! You will insert circular strands of DNA, called DNA plasmids into cells. The DNA plasmids contain genes that genetic engineers have designed to make your cells resistant to a common laboratory antibiotic and to produce a colored pigment.

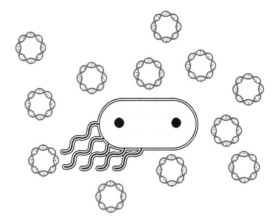

You will build on your mastery from Chapter 3 by growing *E. coli* cells. Then, using chemical processes, you will insert DNA plasmids into your cells so that they read and execute the genes in the DNA plasmid. This will produce the pigment specified by the plasmid.

Step 1.   Download the instruction manual for the Engineer-it Kit

Figure 4-1. Step 1 Download the Engineer-it Kit instructions.

Familiarize yourself with the Engineer-it Kit instructions found at https://amino.bio/instructions. These may reflect updates to the kit's material and instructions. The instructions will also specify what to look for when completing each step. If there are any major conflicts between the most recent manufacturers' instructions and this book, use the instructions as your primary resource.

Step 2.   Put on your gloves and lab coat

Remember to always put on your gloves and lab coat!

Step 3.   Label your plates

Get your Engineer-it Kit from the refrigerator and set up your DNA Playground. Take the **four plates** from the Engineer-it Kit and use a **permanent marker** to label the **bottom** of each plate as follows:

1. **NS; [your initials]**   3. **S; [your initials]; (E)**
2. **S; [your initials]; (+)**   4. **S; [your initials]; (-)**

These abbreviation stand for : NS = non-selective, S = selective, (E) = experimental sample, (+) = positive control sample, (-) = negative control sample

Step 4.   Make non-selective and selective LB agar plates

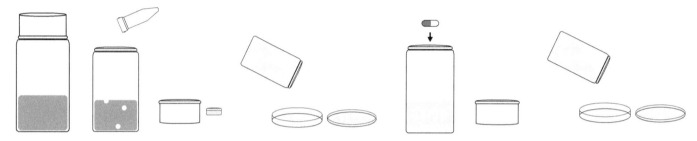

Figure 4-2. Step 4 Prepare your non-selective and selective plates.

Following the manufacturer's instructions, make LB agar plates to grow non-genetically engineered bacteria as well as genetically-engineered bacteria. For this, you will create two kinds of plates: one plate without antibiotics and three plates with antibiotics. Complete the exercise according to the kit instructions and techniques you learned in Chapter 3.

This will be similar to making plates for Chapter 3's Canvas Kit, except this time you will leave one plate without antibiotics. This will be your non-selective plate (NS), which you will use on Day 1, along with your selective S(-) plate. Return the other two selective S(E) and S(+) plates to their resealable bag in and store in a refrigerator until Day 3.

Selective and Non-selective *Going Deeper* **4-1**

Non-selective LB agar plate: This growth plate contains only LB agar, NO antibiotics. They are used to grow bacteria that have not yet been genetically engineered with a selection trait, for example, antibiotic resistance. The K12 *E. coli* bacteria that you grew in Chapter 3 were already engineered with a DNA plasmid to make them colorful and resistant to the antibiotics. This is why in Chapter 3 all of your LB agar plates had antibiotics.

Selective LB agar plates: These plates contain the same LB agar that bacteria need to grow and divide, with something extra added while the agar is still hot and molten. This addition is usually an antibiotic such as ampicillin, carbenicillin, chloramphenicol, kanamycin, or tetracycline. When you engineer your bacteria with the plasmid, the plasmid contains a 'selection marker' – a gene – that results in the creation of a protein that can break down antibiotics. This is called antibiotics resistance, and it makes the bacteria "immune" to the antibiotic. As you will see further in this chapter, this helps you to select for bacteria that you've engineered.

Bubbles in your Agar *Pro-tip*

Genetic Engineering Heroes will often use a flame from a bunsen burner, torch, or a heat gun in order to pop bubbles in their agar plates. To do this, right after you've poured the LB agar into the petri dish, and before it solidifies, you point the heat source at the plate with bubbly agar. The heat causes the air within the bubbles to expand, making the bubbles pop. This is generally only done when creating many plates (20 or more) and it is inevitable to have some bubbles in the agar. To prevent most of them, simply be aware of how vigorously you mix your LB agar powder into the sterile water! **Always be very careful with open flames.**

Making your own LB Agar Plates - LB agar powder *Pro-tip*

You can buy LB agar powder from a supplier to make LB agar plates. Typically, you want to make a ~4% solution of LB agar. This % describes the desired concentration of a solution, typically a measure of weight / volume (w/v) or volume/volume (v/v). In this case, LB agar powder is measured as weight/mass (in grams) and the sterile distilled water is measured in volume (mL). To make a 4% solution, you would add 4 g of LB agar to 100 mL of sterile water.

Concentration (%) = {mass (g) / volume (mL)} x 100

Making your own LB Agar Plates - Antibiotics *Pro-tip*

The Engineer-it Kits include an innovative pre-measured way to add antibiotics to your agar. As you advance as a Genetic Engineering Hero, you may need to create your own antibiotic mixtures. Table 4-1 identifies commonly used antibiotics, the concentrations required to make stock and during experiments.

Note: it is often difficult to get access to 100% ethanol. A replacement solvent is 99% USP isopropyl alcohol (rubbing alcohol) that can be obtained from a pharmacy.

Table 4-1 - Common Antibiotics and Concentrations used for Bacterial Selection				
Antibiotic (abbreviation)	**Working Concentration***	**1000x Stock Concentration****	**Solvent***	**Color Code****
Ampicillin (Amp; A)	100 ug/mL	100 mg/mL (286 mM)	50% ethanol	Orange
Chloramphenicol (Clr; C)	35 ug/mL	35 mg/mL (108 mM)	100% ethanol	Green
Kanamycin (Kan; K)	35 ug/mL	35 mg/mL (74 mM)	distilled water	Red
Tetracycline (Tet; T)	15 ug/mL	15 mg/mL (34 mM)	50% ethanol	Yellow

*Working Concentration is the concentration of the chemical that you want in the actual samples. In the example above, this is the concentration of antibiotic after they are added to the molten LB agar.

**Stock Concentration is the concentration of a master "tube" that you will keep for long-term storage in the freezer. From this stock master tube, you would take a small quantity to add to your LB agar to get to your working concentration. For example, this is similar to the food coloring you would buy in a store. The bottle you get from the store is highly concentrated, and you only need to add a few drops to a large quantity of water to get the color. The important thing to note is that with a stock method, you do not need to make the antibiotic solution from scratch every time.

***Solvent is the liquid that you typically dissolve the antibiotic into to make your stocks.

****Universal color codes are assigned for different antibiotics. Instead of writing the name or abbreviation of an antibiotic on plates, you can use a colored marker to put a notch of the relevant color on the plate. This saves time and still informs you and others what antibiotic is used.

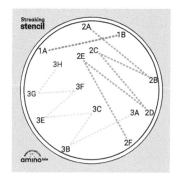

Flour dilution *Practice Breakout Session 2*

The next step of your genetic engineering experiment is to diluting the number of cells you deposit on a plate through a streaking technique involving three inoculation loops. Dilution is the act of reducing the concentration or amount of something. To better understand how the dilution happens through streaking, let's practice dilution with flour.

First, deposit a teaspoon amount of flour (or any similar powder) on a contrasting surface like a wooden table or coloured paper. In this practice, your finger will be your inoculating loop. Have a look at the streaking stencil included in your Engineer-it kit and seen in Figure 4-3; You will want to follow a similar pattern when you dilute your flour. Use your finger to run through the flour pile once, extending the line outwards from the pile. Wipe your finger, and again, trace a line going through the end of your first line, extending outwards. You can even add a few zig-zag, just like on the stencil. Clean your finger again and trace a third and final line that drags through the end of your second line. Extend that third line outwards and add a few zig-zag to further spread the flour. Notice how by the end of your third line, the amount of flour on the surface is much less? You've just successfully diluted flour!

Step 5.   Streaking *E. coli* and the negative control plates

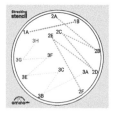 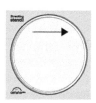 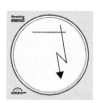 

Figure 4-3. Use the streaking stencil to streak one of your plates with the cells included in the kit.

For this step you will be spreading 'blank' K12 *E. coli* onto two plates: **A.** onto a non-selective (N.S.) plate for use in 12 to 24 hours for the transformation experiment and **B.** on a selective plate S(-) for a negative control sample.

A. You will be using a different streaking method for **K12 *E. coli* bacteria** this time. In Chapter 3, you streaked each color of your bacteria paint with one yellow loop in a zig-zag onto the petri dish with the goal of obtaining lots of growing bacteria. For this exercise with the Engineer-it Kit, your goal is different. Like in the Virtual Bioengineer Engineer-it Kit Edition, you will be streaking onto the **non-selective (NS) plate,** using **three yellow inoculating loops**, following the **streaking stencil pattern**.

This type of streaking is a method used to obtain single colonies of bacteria. A colony is a small mound of bacteria that results from of a single bacterium growing and dividing into millions of bacteria. This means all of the bacteria in that mound are the same.

The goal of this step is to get well-separated colonies of 'Blank Cells'. We call them Blank Cells since they have not yet been engineered - like a blank paper ready for new information. Now that you are initiated into the world of genomic DNA *vs.* DNA plasmids, you will see this is an oversimplification used for beginners; the cells are not entirely blank, but they are not yet engineered.

On your N.S. plate, streak out cells using the stencil and the procedure you learned in Virtual Bioengineer. The stencil included in your Engineer-it Kit has the same multi-colored lines as in Virtual Bioengineer. You should place the stencil under your N.S. plate and follow these instructions (Figure 4-3):

Get a yellow inoculating loop. Dip it into the Blank Cells tube (called a stab of cells). Inspect the loop to see if it looks "wet", indicating that you've dipped into the cells/agar. Then trace over Line 1. Discard the loop in your inactivation bag.

Get a new yellow loop (DO NOT DIP IT IN THE STAB OF CELLS!!!). Follow Line 2 by dragging the new loop through Line 1 and completing the zig zag. Because you are dragging a new loop through a small segment of line 1, you will drag only a small amount of line 1 cells with the loop and spread them out across the rest of the zig zag. There will be far fewer cells in Line 2. Discard the loop.

Repeat with a new yellow loop following Line 3. The third line will have even fewer cells on it. The aim is for you to end up with single bacteria that will grow into mounds called colonies. Discard your loop and place the lid back on the plate.

B. To create your negative control, you will use the simple one-loop zig-zag method from the Canvas Kit exercise. Using **one yellow loop,** you will once again dip into the tube of **Blank Cells** and then drag the loop all over the surface of your **S(-)** plate. Place the lid back on the plate and discard your loop. Learn more about negative controls in the *Going Deeper 4-3* on the next page.

Once you have completed the streaking on your N.S. plate and S(-) plate, invert them and incubate at 37°C for 12 to 24 hours. **Do not incubate for longer than 24 hours before doing the next step or your experiment will not work.**

Streaking *Going Deeper* 4-2

Streaking: It is important to streak the plates to get well-separated colonies because the bacteria in these colonies grow the fastest. The faster the bacteria are growing, the easier DNA can cross their membranes and get into the cytoplasm.

While bacteria grow, they send chemical signals out into the LB agar plate. Some of these signals tell other bacterial colonies to slow down and "stay away from my food", or "Hey! There are a lot of us here, so slow down, don't eat as much". When bacteria receive these signals, they slow their growth and even start bolstering up their membranes in preparation for starvation. By creating different lipids, cholesterol, and proteins, their membranes become rigid and make it harder for you to get DNA into the cells. Well-separated colonies grow much faster because they don't get the "slow down" signals. Their membranes stay more fluid because the cells are dividing fast. This, in turn, makes it easier for you to get DNA into the cells.

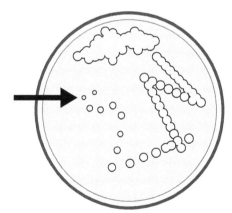

Figure 4-4. Well-separated colonies are found at the "end" of the zig zag dilution pattern, as identified by the arrow.

Negative Control: What is a negative control? In scientific experiments, 'controls' are used to help ensure important components of the experiment – for example, cells, other materials, and cellular processes – are performing as expected. In this exercise, we are going to use controls to help you make sure things are working properly and to help troubleshoot if the experiment doesn't go as expected.

A negative control refers to an experimental sample that you don't expect to respond if it's working properly, hence the word 'negative'. In this negative control exercise, you will be 'plating' (streaking) Blank Cells without antibiotic resistance onto a selective LB agar plate, which has antibiotics. The expected result, if the selective plates were correctly made, is that the Blank Cells should not grow (a negative result). This is because the antibiotics will kill them.

What is the importance of this control for your experiment? This negative control will let you know if the selective plates you need to grow the engineered bacteria in the next steps were made correctly. For example, if you accidentally forgot to add the antibiotics, then the selective plates would not have any antibiotics in them! This would affect your results in a significant way as you would not be able to grow your engineered bacteria optimally. Even if you engineered the cells correctly, you would not likely see engineered cells at the end of the exercise. This control helps you confirm that the selective petri dishes were made correctly, allowing you to have built-in check points to ensure you do not waste materials and time. For example, if your negative control plate has growth on it, you know something is wrong and could make new selective LB agar plates before moving forward with the engineering.

Check Point!

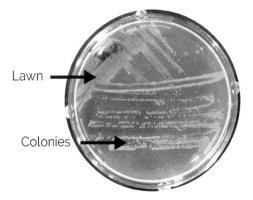

Figure 4-5. Streaked *E. coli* K12 cells, after 16 hours of incubation.

After incubating your Blank Cells for 12 to 24 hours, you should be able to see bacterial colonies growing on your NS plate (Figure 4-5). In the next step, you will be collecting several of these colonies to program them with DNA (insert DNA plasmids into them). Be sure to start Step 6 only after seeing growth similar to what is shown in Figure 4-5 on your own plate. Also, verify that your S(-) negative control plate does not have any growth on it. The S(-) plate tests whether your selective LB agar has enough antibiotics. If your S(-) plate does have Blank Cells growing on it, you will need to make new selective agar plates that are actually selective.

Step 6. Making chemically competent cells

Figure 4-6. Step 6A. Tap the tube of Transformation Buffer on the table to bring all the liquid to the bottom and place it on the Cold Station.

Figure 4-7. Step 6B. Collect Blank Cells - use the well-separated colonies from the NS plate you incubated overnight.

Figure 4-8. Step 6C. Mix your cells in the T. Buffer thoroughly. You should not see any more clumps.

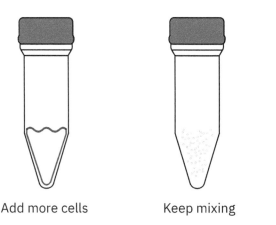

Add more cells   Keep mixing   Cloudy & ready!

Figure 4-9. At the end of Step 6, your T. Buffer should be cloudy with well-suspended cells - no clumps!

We have been discussing inserting DNA into cells in simple terms, but it is important to know that *E. coli* cells do not readily take up new DNA. You will need to use chemistry to "help" the *E. coli* cells take DNA plasmids inside their cytoplasm. Once the DNA is in the cytoplasm, it is "inserted" in the cells. Once you've used chemistry to get your *E. coli* ready to take in DNA, but before you add the DNA, the cells are said to be 'competent'.

A. Tap the tube of **Transformation Buffer** (T. Buffer) on the table to bring all the liquid to the bottom. Then, cool it by placing it on the Cold Station set to 'Ice' on the Minilab. It is important for the transformation buffer to remain cold throughout this process. The salts inside the T. Buffer are harmful and can kill the cells at room temperature, yet are necessary to engineer the cells with new DNA plasmids. Keeping the T. Buffer and the cells that will be added in the next step cold helps to maintain cell stability and survival. To prepare for later steps, turn on the Hot Station to Shock 42 °C so that it warms up.

B. Collect **"Blank Cells"** and mix them into the cold T. Buffer: Using a **small blue 1 uL inoculation loop,** gently scrape the surface of the non-selective LB agar plate and collect some well-separated cells. Colonies that are about 1 mm in diameter or smaller are great! Collect about 10-20 of these colonies so that the center of the loop is full of cells.

C. Mix Blank Cells in cold T. Buffer: Insert the end of the loop with the cells into the cold transformation buffer while keeping the tube on the cold station. Vigorously spin the inoculating loop as though you're making fire! This will cause the cells to fall off the loop and mix in with the cold Transformation Buffer. Mix well so that you see a cloudy solution and no cell clumps (Figure 4-9). This takes about 15 seconds.

At the end of this step, your transformation buffer should be cloudy because of the opaque cells suspended in the clear liquid. If you see clumps of cells floating around you should mix them further by blending vigorously. Imagine it's a cold winter, and you're going to freeze - start the fire!

Once you have a cold Transformation Buffer that is cloudy with cells, your cells have become chemically competent. Your cells are immediately ready for the transformation. **Do not wait more than 3 minutes after adding your cells to the T. Buffer before moving onto the next steps.**

Step 7. Add DNA plasmids and Heat Shock

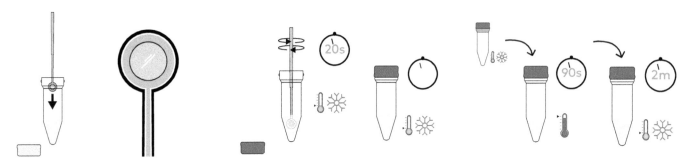

Figure 4-10. Step 7 A. Dip a new blue loop into the DNA tube - make sure you can see liquid in the loop after you dip!

Figure 4-11. Step 7 B. Mix your DNA plasmid in the T. Buffer and cell solution.

Figure 4-12. Step 7 C. Heatshock your cells, T. Buffer and DNA.

Your cold competent cells are ready for you to add DNA plasmids into the mix.

A. Tap the DNA tube on the table so the liquid collects at the bottom of the tube. Dip a **new 1 uL blue loop** in the **tube of DNA**. Swirl and spin the loop for about 10 seconds to ensure the DNA is mixed and that the center of the loop becomes filled with liquid.

Pull the inoculating loop out of the DNA tube and check to see if there is liquid in the center of the loop. Remember blowing soap bubbles? You should see something similar, like the film of soap in a bubble blowing loop. If you can see liquid in the center of the loop, you're ready to add it to your chemically competent cells!

B. Add DNA to your cells in the T. Buffer tube and incubate: Keeping your cells chilled on the Cold Station, dip your inoculating loop with DNA into your chemically competent cells. Swirl and mix in the DNA for 10 seconds. Remove the loop and discard it. Let your cells and DNA remain on cold for 5 minutes. During this time the DNA will begin binding to and interacting with the cells. If you haven't yet, turn the hot station to 42°C, you will need it in the following step.

C. Heatshock the cells: It is now time to get your DNA into the cells. In this step, you will change the temperature of the cells from ice cold (4°C) to hot (42 °C), then back to cold. This is called a heat shock because you are shocking the cells with warm temperature. During the shock, the DNA enters the cells.

Transfer the tube of cells and DNA to the 42°C Hot station. Turn the timer on, wait for 90 seconds. After 90 seconds, return the tube to the Cold Station (4°C). Let it stand for 2 minutes. This traps the DNA inside the cells.

During this short incubation, you can now set the Hot Station to 37°C; you will need this temperature in the next step. The Hot Station will slowly cool down and if it does not reach 37°C by the time you are ready for the next step, that is ok. You can continue onto the steps as it cools down.

What is a plasmid? In the *What is DNA?* you learned that a DNA plasmid is a circular double-stranded helix of DNA. You can imagine DNA plasmids as small DNA molecules that are separate from the cell chromosomal DNA, and that can replicate independently. The DNA plasmid has important sequences and genes that the cell reads once the plasmid enters the cell (Figure 4-13). These are:

Selection marker gene: The Selection marker gene is often used to help select for bacteria that you've engineered while growing the bacteria in LB agar. During the engineering process, a very small number of your competent cells actually take up the DNA plasmid. By giving the engineered bacteria a 'superpower' like antibiotic resistance with the DNA plasmid, you ensure that only your engineered bacteria - cells that have taken in, read and are executing the DNA plasmids successfully – can grow on selective plates with specific antibiotics. Bacteria that didn't take in DNA plasmids (and are therefore not engineered) don't have antibiotic resistance. These cannot survive the antibiotics and die. In simple terms, engineered bacteria grow, and non-engineered bacteria do not.

Origin of replication (ori): The ori is a sequence that is recognized by the cell machinery and tells the cell to copy the DNA Plasmid. This is really important because, without the ori, the plasmid would not get copied and divided as the cells divide! An ori is like the first lines of computer code that specify what libraries and sub-programs (the DNA plasmid) to "include" in the program.

Trait gene: Many different kinds of products can be 'microfactured' by engineered cells. In addition to adding the selection marker gene, another gene is typically added to cause the expression of a new and interesting trait. Usually, this is the trait that tells the cell machinery to produce what the genetic engineer is looking to make. For example, the colored pigment in your Engineer-it Kit.

Negative charges: In Chapter 1, you learned that DNA is negatively charged due to the phosphate ($PO4^-$) molecules that make up its backbone. In Chapter 3, you learned about the lipopolysaccharide (LPS) slime layer and phospholipids (Figure 3-21) that make up the outer surface of *E. coli* bacteria. This outer layer is negatively charged. This is primarily because of the charged head groups of the lipid bilayer. Look back to Figure 3-22 in Chapter 3 and look at the negatively charged group of the example lipid - it is also a phosphate! What happens when two negative charges come into contact? They repel. Now, imagine what might happen if a positively charged ion, like calcium (Ca^{2+}) was present... Your Transformation Buffer is mostly made up of sterile water and Ca^{2+}. One hypothesis as to why T. Buffer helps to get DNA into cells is that the Ca^{2+} ion is able to bind both the negatively charged surface of the cell as well as the negatively charged DNA. In a way, Ca^{2+} acts as a 'glue' that binds to both, causing the DNA to get close to the cell. This is called a 'coordination complex' (Figure 4-14). When the DNA gets cozy with the cells, it increases the chances of the DNA getting into the cells during the heat shock.

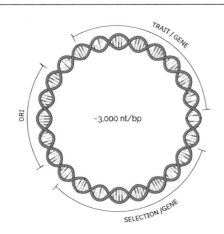

Figure 4-13. A DNA plasmid

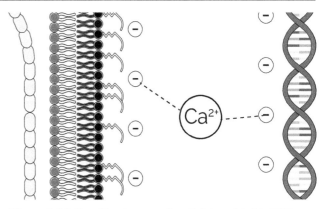

Figure 4-14. Coordination complex. Ca^{2+} ions bind to DNA and the cell's outer membrane, acting like glue between the negatively charged molecules.

Step 8.   Recovery step

Figure 4-15. Pour the Recovery media into the T. Buffer. Mix thoroughly and keep at 37 °C for a minimum of 1 hour, up to 24 hours.

The heat shock step was harmful to the cells, so immediately following the cool down period of the heat shock step, you must 'recover' them.

You will do this by adding **Recovery media** to the cells. Take the lid off both tubes and pour the Recovery Media into the tube of cells and DNA, then close the lid. Note that some liquid will likely stay in the bottom of the Recovery media tube. This is OK, just transfer the majority of the liquid!

Mix the cells in the Recovery media by shaking them. Then, bring all the liquid down to the bottom by either doing the 'Whip-it' maneuver (https://amino.bio/whip-it) or by tapping it on the table.

Place the tube of transformed cells into the Hot Station and set to 37˚C. For the best results, let your cells recover for 12 to 24 hours. This not only allows them to start growing and dividing again, it also gives the cells the opportunity to start expressing the antibiotic resistance gene. This is very important for the next step when you spread your cells on the selective (E) plate that contains antibiotics.

Heatshock & Recovery *Going Deeper* **4-5**

Heat shock: When cells are cold, the lipids in the membrane become more rigid and tightly packed. When they are warmed, the membrane becomes more fluid and permeable. Imagine making bacon on the stove. As you're cooking, a liquid grease comes out of the meat into the pan. After you have completed cooking and the pan cools down, the grease cools and solidifies. This phenomenon may be similar to what happens in the membrane of the cells: the more fluid the membrane, a better chance that DNA can cross the membrane. As you'll see in Chapter 6, this is due to special bonding in the hydrophobic tails of the lipids.

Another analogy is a spa treatment. Often when getting a skin treatment, warm water or steam is used to open up the pores of the skin so that they can be cleaned. After the cleaning, the skin is cooled in order to close the pores. Perhaps a transformation is nothing more than a micro spa treatment - when you heat shock the cells, pores in the membrane form so that DNA can enter into the cells. Upon cooling, the pores contract or disappear and the DNA is trapped inside.

Recovery: Recovery media is LB liquid growth media. The LB agar powder you used to make your petri dishes is similar to the recovery media except LB liquid growth media simply doesn't have any agar. This allows you to grow cells in a liquid broth environment, rather than on a solid gel substance. LB has all of the nutrients and minerals that *E. coli* bacteria need to start growing and dividing again. Just like you probably enjoy a nice meal after some hard work to help you recover your energy, the *E. coli* cells are enjoying LB media. LB is their food source, after the hard work of becoming chemically competent and welcoming DNA into their membrane.

Step 9. Plating and incubating your cells

Figure 4-16. Step 9 A. (left) Plate your recovered cells, DNA and T. Buffer solution. Once it dries, flip before incubation. Step 9 B. (right) Use a yellow loop to streak the Positive cells on your S(+) selective plate.

After your cells have recovered, you can 'plate them'. This simply means that you will pour the cells and recovery media mixture onto one of your selective LB agar plates and spread the fluid across the plate. In this step, you will plate both your experimental cells (e) as well as a positive control (+).

A. Pour ~1/2 of the recovered cells on your experimental selective LB agar plate labeled S(e). Using a new yellow inoculating loop, gently spread the cells over the entire plate and let stand with the lid half-on until the liquid evaporates. This can take up to 30 minutes, depending on your enviroment's temperature and humidity .

If you'd like to speed up the drying process, place the plate on top of the Cold Station of your Minilab with the plate lid half-on. Then turn on the Cold Station. The fan air will aid in evaporating the liquid. After the liquid has fully evaporated, put the lid on, and flip the plate in preparation for incubation with your positive control plate.

B. While your S(e) experiment plate dries, find the tube of **positive control cells (+)** from your Engineer-it Kit. Dip **a new yellow inoculating loop** into this tube of (+) cells and zigzag the loop across the **S(+) control** plate, just like you streaked the negative control. Place the plate lid back on, and flip it in preparation for incubation.

Place your S(e) and S(+) plates into the incubator at 37°C for 24-48 hours. Over this time period, you will start to see bacterial colonies grow. After about 16 hours you will see very small colonies that may or may not be expressing your trait yet (such as a color pigment). At about 24 hours you will see the trait being expressed, while at 48 hours and beyond the trait will become more pronounced. Your (+) control should grow and change color quicker than your newly engineered cells.

Note! In preparation for Chapter 5

Once you see your S(e) plate result, place it in a ziplock style bag in the refrigerator. You will want it in the next chapter's excercise! If you preserved your S(e) plate with a Keep-it Kit, that's ok. A tube of pre-engineered cells will be waiting for you in the kit for Chapter 5. But if you want to use your own engineered cells, don't wait more than one week before starting the Chapter 5 exercise.

Positive Control *Going Deeper* **4-6**

A positive control is an experimental sample in which you expect a known response if the experiment is running nominally. In this experiment, the positive control involves putting previously engineered cells that have antibiotic resistance (+ cells), onto an LB agar plate that contains antibiotics. In this sample, we expect that the engineered cells should grow, as long as the LB agar plates were made properly and have the right amount of antibiotics and nutrients. If the cells do not grow, then this tells us that something may have gone wrong when making the plates!

Step 10.   What to expect & inactivation

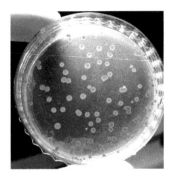
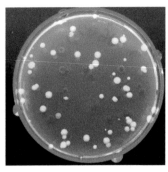

Figure 4-17. Did you get one colony or more on your S(e) plate? Congratulations! One colony or more is a success! The first and last photos are experiments results by *Zero to Genetic Engineering Hero* Junior Editors Pau (first) and Patricia (last)

After 24-48 hours of incubation, if your experiment was successful, you will see colonies (dots) of engineered bacteria. Remember that getting a single colony is a success! Many scientists doing academic or industrial research often hope for a single colony.

If you get more than one engineered colony on your S(e) plate, this means you followed the procedure very well. As you repeat this experiment, you will very likely get more colonies than you did this time, because you will have practiced the procedure and like most things in life, practice makes perfect!

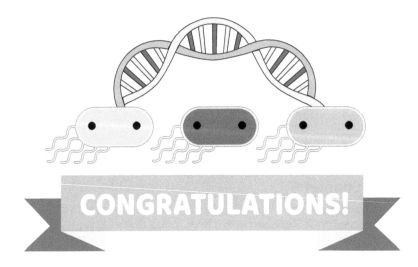

CONGRATULATIONS!

on completing your third experiment - Genetic Engineering!

Fundamentals: How a cell reads a DNA plasmid

The basic operating environment of a cell: The *Four B*'s (Bump, Bind, Burst, Bump)

In a large red brick factory, raw materials are turned into products using machines that rely on workers who pull levers and push buttons, or computers that complete automated protocols through deliberate and intentional actions. Does a cell "microfactory" work like this? How does a cell know what to do?

Cells operate very differently than actual factories. **There are no deliberate or intentional acts by atoms or molecules.** Atoms and molecules do not think or plan out what to do! Rather, three general factors play a role in the 'decision making' or 'logic' of a cell:

- **The number of molecules:** Back in Chapter 1, we learned that what's inside a cell is a bit like a ball pit - it's packed full of molecules. These molecules range from small ones like water to large ones like a cell's genome and protein machinery. Even though this microfactory is very small, it will have thousands or millions of copies of these cellular machines to complete tasks. The quantity of any particular atom or molecule determines how many are available to participate in chemical reactions or other processes. The more molecules or protein machines, the more reactions happen.

- **The rate at which molecules bump into other molecules:** In general, atoms and molecules freely move about the cell bumping into other atoms or molecules at a very fast rate. In the world of the very small, activities happen surprisingly fast. For example, around you right now are trillions of gas molecules – oxygen, hydrogen, nitrogen and carbon dioxide – that fill the room. Do you know how fast they are moving? Hydrogen moves at more than 6,000 kilometers per hour - 60 times faster than a car on a highway! These molecules are so small that we don't feel or notice them.

  Further, the atoms that make up molecules can vibrate at up to 100,000,000,000,000 times per second (10^{14} Hz). The fast movement and vibrations mean that atoms and molecules can bump into other atoms or molecules very often. Within cells, molecules also move fast. Water molecules have an average velocity of 2,000 kilometers/hour.

In other words, atoms and molecules can move and vibrate at high speed over small distances.

Cells are packed full of different molecules which vibrate and bounce around into one another very rapidly. This allows for trillions or quadrillions of interactions to happen in a single cell at any given moment. A single molecule can bounce around and interact with thousands or millions of other molecules every second and, when chemical bonding is strong enough between two molecules, a chemical reaction may occur. Chemical bonding is when atoms or molecules stick together.

- **The strength of chemical bonding between molecules:** Cell 'logic' and decision making is based primarily upon chemical bonding, which is the ability of some molecules to bind specifically or not with other atoms or molecules. This is what causes certain chemical reactions that result in product-making "actions" to take place.

These factors contribute to the basic operating environment of a cell. In short, an extremely large number of "events" involving a large number of atoms and molecules in combination with bonding leads to an action or outcome.

When thinking of the operating environment of a cell, remember the Four B's of Basic Cell Operation:

- **Bump:** Molecules move fast and bump into other molecules in the cell.

- **Bind:** When a molecule bumps into another molecule, it can result in two or more molecules being bound together if the chemical interaction is strong enough.

- **Burst (optional):** In the case of protein enzymes that catalyze chemical reactions, when two molecules interact, a 'burst' or change in energy may occur resulting in a chemical reaction.

- **Bump:** The molecules or products of the chemical reaction separate and continue bumping around the cell.

These are by no means scientific terms, and there are other mechanisms by which certain cellular operations occur, but the *Four B's of Basic Cell Operation* are a great starting point to understanding and remember how *E. coli* cells work. The cells do not have a brain and do not think in the way we understand thinking. Instead, they use these *Four B's.*

Restriction Enzymes *Going Deeper* **4-7**

Restriction enzymes use the Four B's. Restriction enzymes were some of the first genetic engineering tools used by scientists in the 1970s. They are protein enzymes that act like tiny molecular 'scissors' which cut DNA at specific sequences. Restriction enzymes are proteins that have two functions:

1. First, they bind to a specific region of DNA. The restriction enzyme bumps and jostles around the cell, interacting with many other proteins, molecules, and parts of the *E. coli* genome. When a part of the protein, called the binding domain, binds strongly enough with a region of the DNA molecule, the protein will bind like a lock and key due to the width and physical characteristics of the DNA sequence.

2. The second function causes a chemical reaction that can break the chemical bonds of the sugar-phosphate backbone of the DNA's double helix. That means the restriction enzyme can catalyze a chemical reaction that cuts the DNA's backbones. After a restriction enzyme "finds" the appropriate DNA sequence (binding strongly to a region), it cuts the DNA. Once the DNA has been cut the shape of the DNA where the restriction enzyme had bound changes. This change means the restriction enzyme is no longer strongly bound to that DNA. The DNA and enzyme separate, continuing to bump and bounce around until it eventually bumps into another similar DNA sequence elsewhere in the cell.

In other words, the restriction enzyme doesn't think, "I want to find the DNA sequence *GAATTC*." Instead, the unique protein sequence and, therefore, the unique amino acid structures (Chapter 3) can specifically bind to the *GAATTC* DNA sequence. The restriction enzyme bumps around the cell interacting with millions or trillions of other molecules until it interacts with a GAATTC sequence. It then completes its scissor function. It can be argued that all molecules created in the cells have a specific function that the cell has evolved them to have. Combined, these individual functions add up to a very complex machine that seems to have logic.

The *Three Steps to Microfacturing*

Now that you understand the basic rules under which cells operate, let's look at how DNA can be 'recognized,' 'read' and 'executed' by the cell. During the hands-on exercise in this chapter, you genetically engineered your cells with a DNA plasmid. Immediately after you put the DNA into the cells, they began what we call the *Three Steps to Microfacturing.* Microfacturing is like manufacturing, except it happens at the micro-scale or smaller. This is what scientists are referring to when they talk about genetic expression.

We've seen that the order of nucleotides in a DNA sequence make up the 'blueprints' for the cell. The *Three Steps to Microfacturing*, describes three distinct and separate processes that cells use to recognize and read DNA sequences, ultimately decoding it to make cellular products that have a function (Figure 4-18).

- **Step 1 - Transcription:** DNA is a stable chemical molecule with unique sequences. Some regions of DNA have a specific sequence that can be recognized by cellular machinery. As you just learned, this recognition happens when a chemical bond that is strong enough exists between the cellular machinery and the shape and physical characteristics of the DNA segment. In the first step of the *Three Steps to Microfacturing*, DNA becomes bound to and is 'read' by cellular machinery called RNA polymerase. RNA polymerase travels along the DNA reading it and simultaneously transcribing a similar looking, but different nucleic acid string called ribonucleic acid (RNA). Just like how a language translator can listen to Spanish and translate to English in real-time, RNA polymerase reads DNA and simultaneously creates the appropriate RNA molecule. We are going to cover transcription and RNA in depth in this chapter.

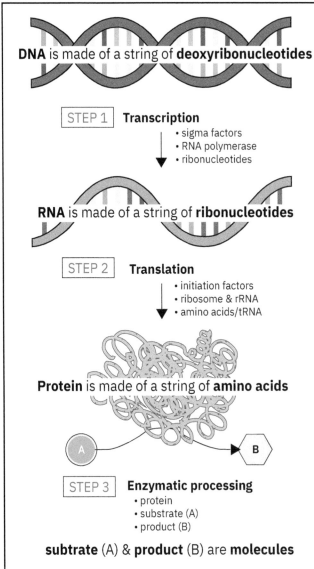

DNA is made of a string of **deoxyribonucleotides**

STEP 1 **Transcription**
• sigma factors
• RNA polymerase
• ribonucleotides

RNA is made of a string of **ribonucleotides**

STEP 2 **Translation**
• initiation factors
• ribosome & rRNA
• amino acids/tRNA

Protein is made of a string of **amino acids**

A

B

STEP 3 **Enzymatic processing**
• protein
• substrate (A)
• product (B)

subtrate (A) & **product** (B) are **molecules**

Figure 4-18. The *Three Steps to Microfacturing* includes DNA being transcribed into RNA, and RNA being translated into proteins. Becoming a Genetic Engineering Hero means knowing how all of these steps work so that you can control what the cell makes, when the cell makes it, and how much it makes!

• **Step 2 - Translation:** As you will learn in the coming sections, RNA is a less stable chemical molecule than DNA, but both molecules share many similar characteristics. In step two of the *Three Steps to Microfacturing*, another kind of protein machinery called a ribosome recognizes a region of the RNA through chemical bonding, along with help from a distinct type of RNA called transfer RNA (tRNA).

The ribosome and tRNA read the RNA, simultaneously translating the RNA sequence into strings of amino acids called proteins. As you learned in

Chapter 3, proteins make up the vast majority of cellular machinery. For example, in this chapter's hands-on activity, the color pigment produced by your cells was likely a protein. The antibiotic resistance was also due to a protein. Translation, tRNA and protein creations will be covered in Chapter 5.

• **Step 3 - Enzymatic Processing:** The third step of the *Three steps to Microfacturing* is Enzymatic Processing. This includes how protein enzymes cause the chemical reactions that make life happen. We will cover it in Chapter 6.

Now, let's have a deeper look at Step 1, transcription. To do so, we will first learn about a very important nucleic acid, ribonucleic acid, or RNA.

Deoxyribonucleic acid (DNA) vs. Ribonucleic acid (RNA)

During transcription, cell machinery reads DNA and simultaneously creates a "sister molecule" called RNA. What is RNA?

An RNA strand looks a lot like a single strand of DNA. The main difference is that the nucleotides making up RNA are very slightly different from DNA nucleotides. Look at Figure 4-19. Can you spot the difference?

If you closely compare the ribose sugar ring of the nucleotides, you'll notice that there is a slight difference at the "bottom" of the ring. The ribonucleotide (RNA) has two "OH" groups - one on the C3 carbon atom, and one on the C2 carbon atom. The DNA's deoxyribonucleotide has only a single 'OH' group, with the C2 carbon instead having a hydrogen 'H'. Deoxy, the removal of the 'oxy' or oxygen atom at the C2 position, is why DNA is called deoxyribonucleic acid. RNA, on the other hand, has two OH groups on the ribose sugar and is referred to as ribonucleic acid.

Beyond this slight difference in the ribose sugar ring, the ribonucleotides of RNA connect together in the same way as DNA (Figure 1-17). The OH on C3 of one nucleotide ribose sugar connects to the phosphate on C5 of another, and this repeats to create a sugar-phosphate backbone. Notice that when two nucleotides are connected, the "H" on the "OH" is removed during the reaction (Figure 4-20).

A second difference between DNA and RNA is that RNA does not form a double helix structure like DNA. As you'll see at the end of this chapter, RNA

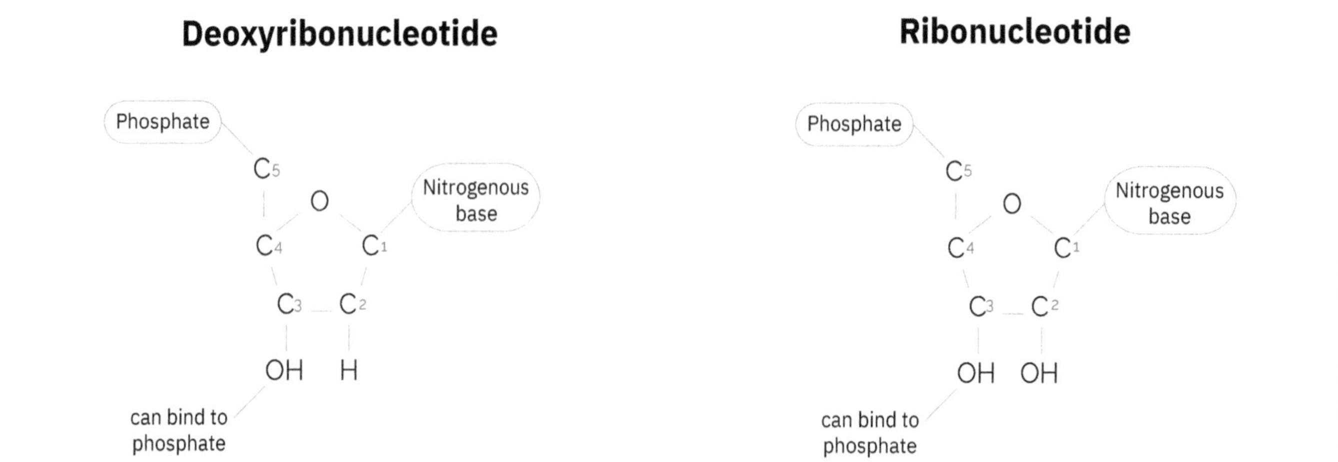

Deoxyribonucleotide

Phosphate

C5

O

Nitrogenous base

C4   C1

C3   C2

OH   H

can bind to phosphate

Ribonucleotide

Phosphate

C5

O

Nitrogenous base

C4   C1

C3   C2

OH   OH

can bind to phosphate

Figure 4-19. Comparing the nucleotide of DNA (left) and RNA (right), you'll notice only a very minor difference in the structure.

nucleotides can complement and bind to one another and form structures, but this happens at a much lower frequency than DNA. This is why in Figure 4-18, the RNA is a lone single strand, while DNA is illustrated as a double helix.

This slight change in the ribose sugar and the fact that it doesn't broadly form a double helix with complementary RNA has profound effects on RNA's stability and function. DNA is very stable. So much so that it can remain intact for millions of years. RNA, on the other hand, is not very stable. Once the cell creates RNA, it stays intact for only a short time before falling apart in minutes or hours.

A four-nucleotide string of RNA can be found in Figure 4-20. You'll see it has a very similar structure to a single strand of DNA. One difference is that the RNA nucleotide has the extra "OH" hydroxyl group in the ribose sugar on carbon C2. Another difference is that RNA does not have thymine. Instead, it has "uracil" (U) nitrogenous base in the uridine ribonucleotide.

Now that we know more about the structure of RNA, we can learn about transcription, the process that the cell uses to read DNA and transcribe it into RNA. To do this, we are going to look at what a gene is and ask these questions:

- What is the cell machine that catalyzes the chemical reaction during transcription?

- How does it know where to start transcribing?

- How does it know what to transcribe?

- How does it know when to stop transcribing?

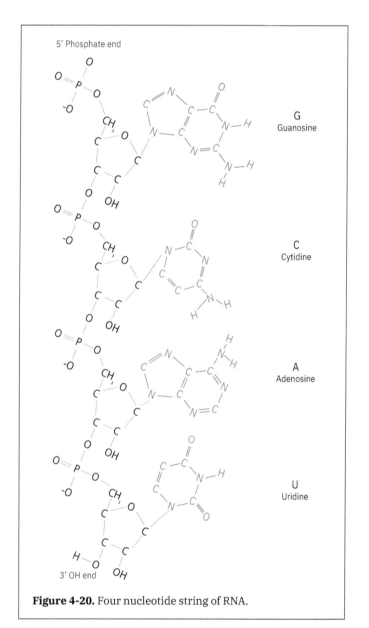

Figure 4-20. Four nucleotide string of RNA.

RNA polymerase: The cell machine that transcribes

The cellular machine responsible for transcription is a protein enzyme called RNA polymerase.

RNA: because it catalyzes the creation of RNA

Polymerase: because it joins RNA nucleotides (ribonucleotides) together into a string. Polymer is a general chemistry term to describe a chemical string of one or more reoccurring building blocks.

Just as RNA polymerase is involved in expressing the genes that were produced by your engineered *E. coli*, RNA polymerase itself is expressed from a gene in the genome of the cell. That means RNA polymerase is involved in creating itself!

As you'll see later in this chapter, RNA polymerase is able to bind to a DNA strand (Figure 4-21). It can then ride along the DNA like a train on tracks, and simultaneously create (transcribe) an RNA molecule.

Compared to the width of the DNA strand, RNA polymerase is much wider and is able to surround the strand of DNA. In Figure 4-22, the RNA polymerase (pink) surrounds a piece of DNA (orange/blue). Unlike the artist's depiction in Figure 4-21, Figure 4-22 is a mathematical model based on real data of RNA polymerase bound to DNA. So while you don't see a full DNA strand passing through the RNA polymerase, this illustration provides a very real perspective on the size of RNA polymerase compared to DNA.

In coming sections, we will go into much deeper detail about how RNA polymerase knows where and when to start and stop transcribing.

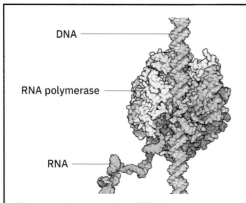

Figure 4-21. Artist view of RNA polymerase protein (blue green) bound to a DNA strand (orange) and transcribing RNA (red). Source: Protein Data Bank (PDB) Bruce Alberts, A.Johnson, J. Lewis, M. Raff, K. Roberts and P. Walter (2002) "Molecular Biology of the Cell" Ch. 6, Garland, New York.

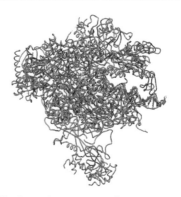

Figure 4-22. Crystal structure of RNA polymerase (pink) bound to a DNA helix (orange/blue). When operating within the cell, the DNA continues through RNA polymerase, and an RNA strand protrudes out the surface. Source: Ibid

What is a gene?

A gene is one of the most talked about, but least understood, topics in education. A gene is a length of DNA that has all the DNA sequence the cell needs to read and begin the *Three Steps to Microfacturing*. In other words, a gene is a length of DNA that results in the creation of an end-product that has a function, like RNA or a protein.

This means a gene must have information embedded in the DNA sequence to start and stop the *Three Steps to Microfacturing* and to create a cellular product with a function. In other words, a gene is a length of DNA that can tell the cell machinery (RNA polymerase) when and where to start the *Three Steps to Microfacturing* as well as what to make. Let's look deeper at how it works.

Just like a sentence has a structure or "syntax", such as a subject-verb-object, genes have a grammatical order. Two kinds of information are stored in a gene's DNA sequences. They are called "non-coding sequences" and "coding sequences" (Figure 4-23). These are both just plain old DNA. The RNA polymerase is able to distinguish between them.

A non-coding DNA sequence is a segment of DNA that acts as a switch, controlling when and how much product is made from the gene. The non-coding sequence has the right characteristics to bind 'transcription machinery', and it acts as the starting point of transcription. Consider The Four B's of Cell Operation: Bump, Bind, Burst, Bump. If the non-coding DNA sequence is unable to bind the transcription machinery, then transcription doesn't happen. Conversely, if the DNA sequence has the right shape and charge to bond to the transcription machinery, RNA polymerase binds to the DNA more frequently, and transcription can occur.

In the hands-on exercise, you engineered your cells by adding a DNA plasmid that contains a gene for creating protein color pigments. Within that gene is a non-coding sequence designed to bind with a cell's transcription machinery ~12 hours after the cells start growing and keep transcribing it thereafter.

A coding DNA sequence is a sequence situated immediately next to the non-coding DNA. It is read and transcribed by the transcription machinery into RNA. The coding DNA sequences are like the designs for the functional end-product that will be made from the gene. The non-coding DNA sequences are like the switch telling the cell where and how frequently to transcribe the coding DNA sequence.

Starting Transcription

How does the cell know how to start transcription? Both the DNA and transcription machinery are bumping around the cell. If the non-coding DNA sequence has the 'right' shape and chemical bonding properties, it will bind to the transcription machinery, enabling transcription to start. Let's get more specific.

Small proteins called sigma factors complete the Four B's and eventually bind to short non-coding DNA sequences within genes. For transcription, the small non-coding regions are called promoters (Figure 4-23) because they "promote" the transcription of the gene. Promoters are the starting points of transcription, which means they are the starting point of a gene. The sigma factor (σ) has a particular size and shape that is able to bind to a specific DNA sequence. Once a sigma factor binds to the promoter of a gene, it then binds to the RNA polymerase. In other words, the sigma factor acts as a bridge between the DNA and the RNA polymerase, the enzyme that transcribes RNA (4-23i).

Once RNA polymerase is bound to the promoter region via the sigma factor, the RNA polymerase creates a short RNA sequence called the initiation sequence, which locks the RNA polymerase to the DNA (4-23ii). The RNA polymerase then "drives off" and escapes the promoter to begin unzipping, reading, and transcribing the coding DNA sequence into RNA (4-23iii).

An analogy can be drawn between transcription and drag car racing (do a quick video search for 'drag car racing'). Note the similarity:

- **Step 1:** A race track official walks onto the starting line (a sigma factor binds to the promoter, which is the 'start line' of the gene).

- **Step 2:** The official waves the car forward and the car advances to the starting line (the sigma factor 'recruits' and binds to RNA polymerase at the promoter).

- **Step 3:** The car does a 'burn out', spinning its tires to heat them up, making the tires nice and sticky, so it has more traction (the RNA polymerase makes a short RNA initiation sequence, locking into the DNA).

- **Step 4:** Green light! The race car floors it and takes off from the starting line (RNA polymerase leaves the promoter and commences transcription).

Sometimes the car makes a perfect escape. However, in some cases, the race car does a wheelie, burns out, or the engine fails, and the car cannot effectively escape the starting line. Often when RNA polymerase attempts to start transcription, it cannot complete all of these steps and fails to begin transcription.

This is called abortive initiation. If in the future you want to design your own promoter DNA sequences, keep in mind that controlling transcription is a fine balance. You want the RNA polymerase to bind to the promoter, but not too tightly or it won't be able to escape the promoter!

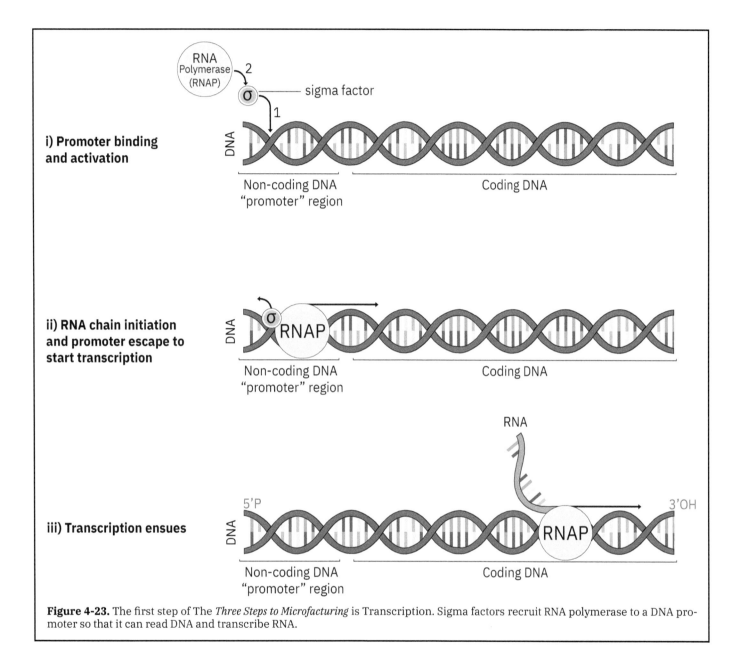

Figure 4-23. The first step of The *Three Steps to Microfacturing* is Transcription. Sigma factors recruit RNA polymerase to a DNA promoter so that it can read DNA and transcribe RNA.

Transcription *Video Breakout*

You can search the web for some amazing videos that include computer graphic images of RNA polymerase doing the Four B's: Floating around, recognizing/binding to a non-coding DNA sequence, and then riding along and unzipping the coding DNA sequence (reading it), all while transcribing an RNA molecule. Search terms could include: "RNA polymerase video", "DNA transcription video".

Know your strand! *Breakout Exercise*

You've learned a lot of new information in the last few pages. Take a break to reflect on what you know about DNA by completing this short exercise.

Label the strand with the items below.

- **5' Phosphate**
- **3' OH**
- **deoxyribophosphate backbone**
- **base pairs**

- **hydrogen bond**
- **hydrophilic**
- **ionic bond**

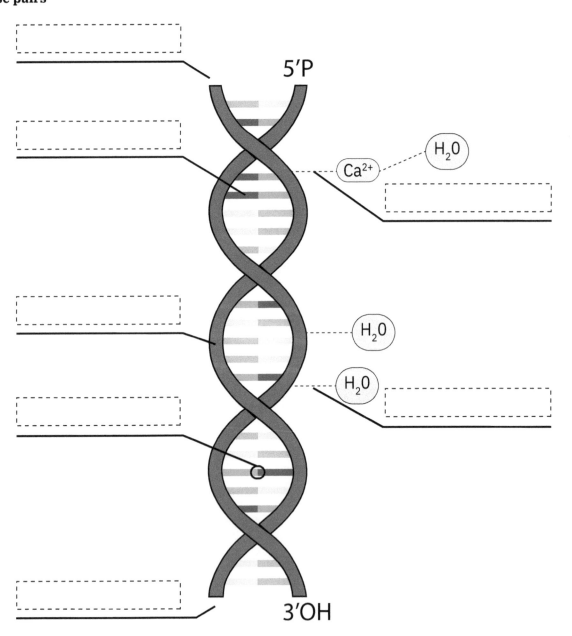

Don't forget! If you are stuck or want to verify your work, go to www.amino.bio/community

Are there more than one type of sigma factor? Yes, there are many! The non-coding regions of genes have co-evolved with many different sigma factors so that cells can have many different switches to turn genes on and off. These enable the cells to create products at different times, under different growth conditions, and within different environments! This enables cells to create certain products only when they are needed.

For example, a sigma factor called sigmaS (σS) binds to certain promoter sequences that control the transcription of proteins that are important when cells are under starvation and are preparing to stop growing, getting ready for 'tough times'. When cells are growing fast, very little sigmaS is created by the cell. However, as food becomes limited or the cells become crowded, the cells produce more sigmaS, which then can bind to the non-coding promoters of genes and recruit RNA polymerase. This activates transcription resulting in RNA that will later be translated into important proteins which ready the cell for 'hibernation'!

Going Even Deeper

You may have just realized that sigmaS is a protein, meaning its creation through the *Three Steps to Microfacturing* must also be controlled by a sigma factor... and you're right, it is! The coding DNA for sigmaS is controlled by a promoter called rpoSp, which is also able to bind to another sigma factor. It gets more complicated quite quickly, and it's not necessary to dive deeper right now. However, if you're interested, you can do a web search on this topic.

In the context of the hands-on experiment you completed in this chapter, the gene for the color pigment is controlled by a sigma factor that turns on when the cells enter a "stationary phase".

During transcription: Direction

Just like roads have lanes that operate in different directions, DNA and RNA have direction! Let's get oriented on how directions work in nucleic acids. Recall that DNA and RNA nucleotides have a ribose sugar that is bound to an OH group at carbon C3 and a phosphate group that is bound to carbon C5 (Figure 4-19). It is common language to say that the phosphate attached at the C5 carbon position of the ribose sugar is the "5' phosphate" ('five-prime-phosphate'). The OH group on carbon C3 is referred to as the "3' OH" ('three-prime-o-H'). The apostrophe is pronounced as "prime".

These two positions of the ribose are what become connected to form a chain of nucleotides. If you look again at Figure 1-17, you will see that at the beginning of the strands there is a 5' phosphate group that is not attached to anything. Also, you'll see at the other end a 3' OH that is not connected to anything. These two end groups are how we understand the position and directionality of DNA.

In the world of nucleic acids, the 5' phosphate of a DNA (Figure 1-17) or RNA (Figure 4-20) strand is always considered the 'beginning'. The 3' OH of the DNA or RNA is considered the 'end' and cellular machines such as RNA polymerase, travel from the 5' P toward the 3' OH (Figure 4-20). A very common way in which scientists describe the location in a DNA sequence is using 'upstream' and 'downstream'. The 5' phosphate is upstream of the 3' OH (Figure 4-24).

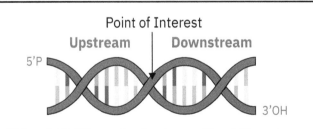

Figure 4-24. When using "upstream of" and "downstream from" terminology, you refer to whether something is closer to the 5' phosphate (upstream) or the 3'OH (downstream) of a specific strand. Be careful! You also have to know what strand you're referring to. Here, we are referring to the pink DNA strand.

Genetic Engineering Heroes often use upstream and downstream to describe a location or direction on a strand of nucleic acid. For example, you can say "the promoter is just upstream of the coding region", or "I get it, the coding sequence is just downstream of the promoter!" (Figure 4-24).

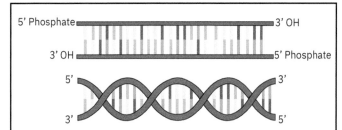

Figure 4-25. Strands of nucleic acids have directionality, meaning one end is different from the other. When two strands of DNA bind together, they bind in an anti-parallel manner so that the 5' end of one strand is at the 3' end of the other.

But DNA has two strands. What happens when they come together? There are two directions, and each depends on which strand you are referring to. When two DNA strands zip together to form the DNA double helix, the strands bind in an "antiparallel fashion", meaning that the 5' phosphate ends of each strand are at the opposite ends (Figure 4-25). Look back to Figure 1-17 and notice where the 5' phosphate and 3' OH are located on the two strands. This means that the 'upstream' and 'downstream' terminology depends on which strand you're referring to.

During transcription: Which DNA strand does RNA polymerase read?

There are lots of nucleic acid strands to keep track of! During transcription there are three nucleotide strands involved:

- The RNA strand that is being made (transcribed) by RNA polymerase

- The two complementary DNA strands that make up the DNA double helix, only one of which is being "read" by RNA polymerase

How does RNA polymerase know which strand of DNA to bind to so that it goes in the right direction?

The RNA polymerase binds to both strands simultaneously, but the direction it is pointing depends on the strand that the promoter sequence is in. In Figure 4-27 you'll see that the promoter (gray region) can actually be situated on either strand, with the arrows pointing in the downstream direction indicating the direction that the RNA polymerase would travel. It is a specific DNA sequence in the promoter which helps lock the sigma factor and RNA polymerase in the right direction (Figure 4-26). The DNA sequence ATCG's cause the DNA helix to have slightly different shapes

that the sigma factor can bind to in a specific orientation. The orientation can help determine which direction the polymerase points.

Whichever strand the promoter sequence is situated in is called the 'plus' (+) strand or 'leading strand' and the RNA polymerase will be oriented so that it will transcribe downstream of the promoter into the coding region (Figure 4-26).

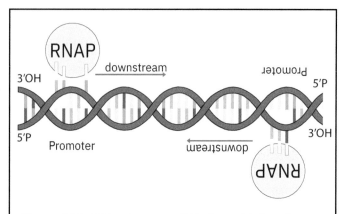

Figure 4-26. RNA polymerase (RNAP) can bind to the DNA nucleotides in one way which points the RNA polymerase in the correct direction.

Here's where things get a little wacky - ready for a mind bender? Even though our point of reference is the (+) strand, and the RNA polymerase travels from 5' to 3' according to the (+) strand, the RNA polymerase actually reads and transcribes from the other strand, the template strand. Another name for this is the 'minus' (-) strand. In the next section, we're going to see why RNA polymerase does this, along with the 'cipher' that it uses to transcribe DNA sequences into RNA sequences.

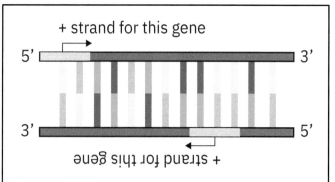

Figure 4-27. The gray segment of each strand represents different non-coding promoters. On the top strand, the RNA polymerase would bind and continue to the right toward the 3' OH end. On the bottom strand, the RNA polymerase would bind to the gray region and transcribe to the left toward the 3' OH end of that strand.

How does RNA Polymerase know which direction to go? *Going Deeper* **4-9**

You just learned that DNA is non-symmetrical and has "directionality". This means that the ends of the DNA strands are different. You also learned that RNA polymerase travels in one direction, from 5' phosphate downstream to the 3' OH. So how does the RNA polymerase know which way to go? A simple answer is that RNA polymerase has evolved to go in one direction. Similar to how gears can be designed to turn in one direction (see: https://amino.bio/pages/ratchet-gear), RNA polymerase operates only in the 5' phosphate to 3' OH direction.

But how does the RNA polymerase point toward the 3' OH so it does travel in the right direction? Figure 4-26 shows how some nucleotides can interact with the surface of the RNA polymerase and cause it to point in a direction. While this illustration is a simplified version of what happens in real life, you'll see that there is only one orientation in which the RNA polymerase can bind, and this will point the RNA polymerase in the correct direction.

Bidirectional transcription in a plasmid? *Going Deeper* **4-10**

Have a look at Figure 4-28. This is an adapted illustration of Figure 4-13 where the twist of the DNA helix was removed to show each DNA strand more clearly.

- Your trait/gene: is designed so the RNA polymerase will bind to the promoter (5'P) and transcribe towards the 3'OH direction of the pink (outer) strand. While this happens, it uses the red (inner) strand as a template to create an RNA molecule version of the pink strand.

- Selection gene: is designed so the RNA polymerase will bind to the promoter (5'P) and transcribe towards the 3'OH direction of the red (inner) strand. While this happens, it uses the pink (outer) strand as a template to create an RNA molecule version of the red strand.

The Ori is not involved in transcription. This bidirectional design is often used so that the RNA polymerase from one gene does not continue into the next gene and transcribe it as well. This could decrease your control of the genetic system!

In the figure on the left, even if the RNA polymerase continues past your trait gene, it will not transcribe the selection gene because that RNA polymerase is reading the wrong strand. On the right it is possible that transcription starting in a gene can run through another gene!

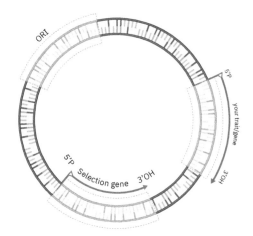 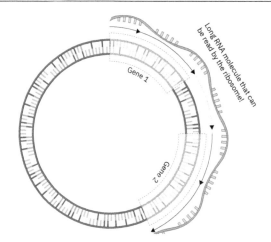

Figure 4-28. Left: Bidirectional transcription in a plasmid.   **Right:** Unidirectional transcription in a plasmid.

During transcription: A secret cipher for transcribing DNA to RNA

After discovering DNA's structured in the 1950s, the next great mystery was how DNA could hold the information for other molecules such as RNA and proteins. It took decades of research to unravel this.

As you saw in Figure 4-19 and Figure 4-20, RNA nucleotide structures are quite similar to DNA nucleotides. There is one other significant difference between DNA and RNA: There is no 'T' thymidine nucleotide in RNA. Instead, there is a 'U' nucleotide for uridine, hence, GCAU.

In Chapter 1, you saw that certain nucleotides can complement each other using Chargaff's Rule. Notably, A's can bind to T's to form a double-stranded DNA, and C's can bind to G's. This complementary rule is how RNA polymerase knows which ribonucleotides to match up with when transcribing an RNA strand, and this is why it always reads the template strand. It uses the template strand because it is a 'mirror image' of the leading (+) strand. The end goal of RNA polymerase is to have an RNA strand that is a replica of the leading (+) DNA sequence, making the (-) template strand its 'mirrored strand'.

Lastly, because there is no 'T' in RNA, T's are replaced with U's, (Table 4-2). This means that the 'A' in a (-) template strand interacts with a 'U' instead of a 'T'. In other words, if there is a 'T' in your DNA (+) strand, then there will be a 'U' in the RNA strand.

As RNA polymerase moves downstream reading the template strand of the DNA, ribonucleotides doing the Four B's bump into the RNA polymerase. When the correct complementary ribonucleotide (Table 4-2) bumps the correct DNA nucleotide inside the RNA polymerase (e.g., A-U or G-C or T-A) it will bind and trigger a chemical reaction burst, permanently attaching itself to the growing string of RNA (Figure 4-29).

If an incorrect match occurs (e.g., A-G), then the ribonucleotide will bump out. Eventually, a correct ribonucleotide will bump in, bind strongly, and trigger the reaction. When this chemical reaction happens, it also propels the RNA polymerase to move further down the DNA strand to the next nucleotide. The RNA polymerase moves at about 50 ribonucleotides per second - that's fast! Can you do the following exercise just as fast? Complete Table 4-3 in less than one second!

Be the cell machinery! *Breakout Exercise*

There's been a lot to take in! Have a look at Table 4-2 and be the RNA polymerase! Complete Table 4-3.

Table 4-2. Transcription cipher - Nucleotide pairing table		
DNA + leading strand	**DNA- template strand**	**RNA**
A	T	A
T	A	U
C	G	C
G	C	G

Table 4-3. Be the RNA Polymerase		

5'   3'

Leading DNA (+)	c a t g c g t g c a a a a c c c a t g a a c c g c t g g c g a a c g a a a c c
Template DNA (-)	
RNA	

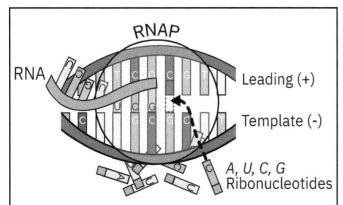

Figure 4-29. As RNA polymerase travels downstream unzipping the double-stranded DNA, it "reads" the (-) template strand. Free-floating, complimentary ribonucleotides (C-G or U-A or T-A) will bind strongly to the nucleotide of the DNA's template strand within the polymerase. This triggers the polymerase to attach the ribonucleotide to the growing RNA strand. If a nucleotide doesn't bind strongly (for example a C-A), it will bump out of the RNA polymerase. Eventually, the correct one will enter and be added to the growing string of RNA.

Stopping transcription

How does the RNA polymerase know when to stop transcribing a gene? In *E. coli*, RNA polymerase will stop transcribing a DNA sequence in one of three ways:

- **The RNA polymerase 'slips off' the DNA:** If you look back to Chapter 1, where you learned about the structure of double-stranded DNA and the zippering that occurs between complementary nucleotides, you'll notice in Figure 1-17 that there are a different number of bonds (dashed lines) between a G-C pair compared to an A-T pair. A 'G-C' complement has three bonds that hold the complementary nucleotides together. An 'A-T' pair, has only two bonds. More bonds mean stronger interactions, which means that the bonding strength between A-T is weaker than G-C.

When RNA polymerase is riding along transcribing DNA, it uses the bonds between the transcribed RNA and the DNA (-) template strand to hold itself connected to the DNA. A long stretch of repeat T's in the DNA's (+) leading strand (which correspond to A's in the (-) template strand) results in a string of U's in RNA (U-U-U-U-U...), each of which also only has two bonds. This results in weak interactions between the RNA polymerase and the DNA, and often the RNA polymerase simply slips off. You will often find stretches of T's in DNA (U's in RNA) at the end of a gene. These are placed to cause the RNA polymerase to fall off of the DNA and stop transcription.

- **The RNA strand folds up, causing RNA polymerase to fall off - a 'terminator':** What would happen if ribonucleotides of an RNA string were able to interact with other ribonucleotides in the same string? You've seen that two different strands of DNA can come together to form a double helix. Can something similar happen with RNA? Yes!

Because RNA transcripts are quite flexible, they can flip and flop around, allowing nucleotides of the RNA strand to come into contact with one another. A string of RNA can interact with itself, and similar rules apply: A binds to U, G binds to C.

When this happens, a 'hairpin structure' can form. In the example in Figure 4-30, you'll see what is called the bacteriophage 82 late gene terminator. Two regions of the RNA transcript complement well enough to form what is called a stem and loop. The overall structure is called a hairpin, and the formation of the hairpin structure creates physical stress between the RNA transcript in the RNA polymerase. This causes the RNA polymerase to fall off of the DNA. In most cases, there is also a poly-uridine, also called poly-U, segment (U-U-U-U-U...) immediately following the stem. This results in the RNA polymerase slipping off of the DNA, as you saw in the first example.

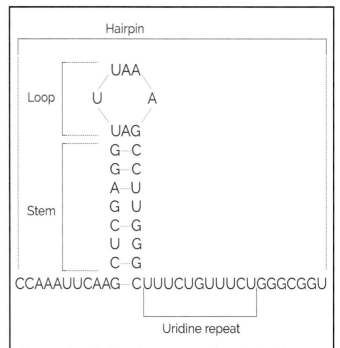

Figure 4-30. A hairpin is a structure where RNA folds upon itself to create a structure that causes the RNA polymerase to get jammed up and detach from the DNA. Another name for this is a 'terminator'.

- **Another protein chases RNA polymerase off the strand:** Rho is a protein that has the function of actively stopping transcription from happening. The way Rho works is really cool. As RNA polymerase continues to move downstream and transcribe RNA from DNA, Rho is able to bind near the 5' end of the new RNA transcript and move downstream on the strand toward the 3' end as if it is chasing the RNA polymerase (Figure 4-31). The Rho protein eventually catches up to the RNA polymerase and is thought to tug the RNA strand out of the RNA polymerase, resulting in the termination of that transcript.

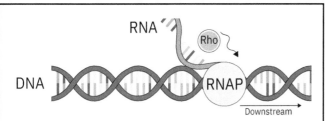

Figure 4-31. A protein called Rho is able to bind to RNA transcripts. As RNA polymerase is transcribing, Rho will glide up the transcript. When it reaches the RNA polymerase, it can 'tug' the RNA out of the polymerase, which stops transcription.

What can you do with RNA?

You've now learned what it takes to start, do, and stop transcription! So, what can you do with RNA?

RNA has many known functions, and many more will be discovered in the coming decades. The most understood use of RNA is the subject of Chapter 5, where we will look at how RNA is read by cellular machinery and translated into proteins. Proteins make up the vast majority of cellular machinery that catalyze chemical reactions or form cell structures.

RNA itself can also cause chemical reactions. 'Ribozymes' are short strands of RNA that fold in the right way so they can cause chemical reactions to happen. Maybe you have heard of CRISPR-Cas9? An integral part of the CRISPR system working includes using RNA to guide a protein to a specific DNA sequence.

There are many more functions and uses for RNA, and you are now well-equipped to explore further! This is a topic of great interest amongst genetic engineers and you will undoubtedly be able to discover many resources by exploring the web.

What is life without RNA polymerase? *Web Search Breakout*

As you now know, RNA polymerase is an extremely important enzyme that is essential for life. What would happen if RNA polymerase couldn't function in your cells?

If you look into nature, you might run into a very poisonous mushroom called the "death cap". This mushroom looks like an ordinary mushroom but has a slight green color. The death cap mushroom naturally microfactures a molecule called an amatoxin. Amatoxin is able to bind to RNA polymerase and sllllooooooooooww down its function. Rather than being able to transcribe thousands of ribonucleotides per minute, RNA polymerase can only assemble a few ribonucleotides per minute. This means that cells with amatoxin in them have a hard time completing the *Three Steps to Microfacturing* - which is a big problem for cell operation.

Visit amino.bio/mushroom for more information about the death cap mushroom.

Be the cell machinery! Bidirectionally! *Breakout Exercise*

You have recently learned that RNA polymerase can transcribe from both strands of a DNA helix. In Figure 4-28 of Going Deeper 4-10, you can see that genetic engineers will design their DNA in a way that genes will be transcribed in opposite directions from differing strands in order to prevent one gene from reading into the next.

In this Breakout Exercise, you will find a DNA sequence across the middle of the table. Your goal is to be an RNA polymerase and correctly transcribe the DNA sequence using the knowledge you've learned. Some tips include:

- keep an eye on the 5' phosphate and 3' hydroxyl ends of the DNA as these will help you know which direction to transcribe
- recall what an RNA polymerase transcribes from DNA (*e.g.* does it transcribe the promoter region?)
- recall which strand the polymerase "reads" in order to create the RNA transcript (*e.g.* the (+) or (–) strand?), and that the (+) and (–) strand designations depend on which strand you're reading
- recall the DNA to RNA cipher

A very important take-away from this exercise is that the RNA transcripts that are generated from each DNA strand are unique! This means that the cell can create different RNA from the same strand of DNA if it is read in opposite directions!

You'll also see that there is space available to complete the cell process called translation (the protein rows). This is covered in Chapter 5, and you'll be able to come back and finish this table once you've learned more about how the ribosome is able to "read" an RNA transcript and create a chain of amino acids, also called a protein.

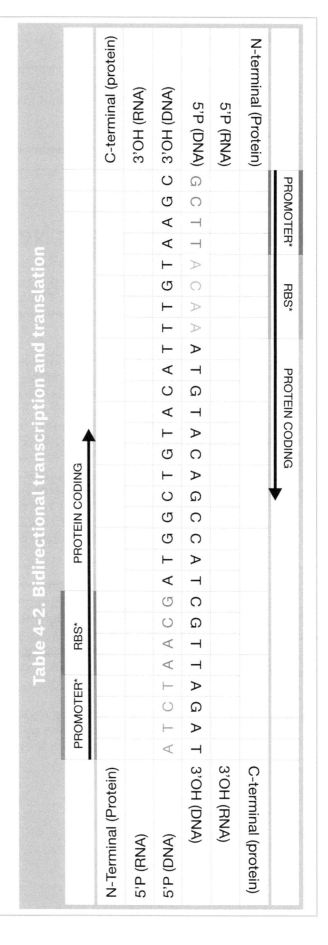

Table 4-2. Bidirectional transcription and translation

	PROMOTER*	RBS*	PROTEIN CODING	
N-terminal (Protein)				C-terminal (protein)
5'P (RNA)				3'OH (RNA)
5'P (DNA)	G C T T A C A A A T G T A C A G C C A T			3'OH (DNA)
3'OH (DNA)	A T C T A A C G A T G G C T G T A C A T T T G T A A G C			5'P (DNA)
			T A G A T	
3'OH (RNA)				5'P (RNA)
C-terminal (protein)				N-Terminal (Protein)

Summary and What's Next?

In Chapter 4, you explored the Basic Operating Conditions of a Cell, as well as the first step in the *Three Steps to Microfacturing*: Transcription. Here is a summary of the transcription process:

1. A sigma factor (protein) bumping around in the cell binds to the non-coding promoter of the DNA (+) strand. Because there are two strands that form the DNA helix, the (+) strand can be either strand as decided by the genetic engineer and therefore the RNA polymerase can point in either direction.

2. Because the sigma factor is able to bind to both the DNA promoter and the RNA polymerase, once attached, the sigma factor 'recruits' the rest of the RNA polymerase machinery to bind to the DNA. There are many copies of RNA polymerase in the cell at any given moment.

3. With the template strand, RNA polymerase reads a short piece of the DNA and transcribes an initiation RNA sequence - locking the RNA polymerase onto the DNA sequence due to the A-U or C-G or T-A bonds. The RNA polymerase then escapes the promoter and begins unzipping the DNA (separating the two strands).

4. The RNA polymerase reads the coding region of the DNA (-) strand and by matching RNA nucleotides to it. When the correct ribonucleotide enters the RNA polymerase and matches the DNA nucleotide (A-U, C-G, T-A), the ribonucleotide is added the growing string of RNA. The RNA will have a similar sequence to the DNA sequence (+ strand), except any T's are instead U's. As the RNA polymerase creates the string, some of the chemical reaction energy from adding each new ribonucleotide propels the RNA polymerase downstream.

5. Using the promoter in the leading (+) DNA strand as a reference point, the RNA polymerase moves downstream from the 5' end of the leading (+) DNA strand toward the 3' end of the leading (+) DNA strand. This also means that the RNA polymerase creates an RNA transcript from the 5' phosphate end to the 3' OH end.

6. Due to slippage, a terminator, or proteins called Rho, the RNA polymerase falls off of the DNA. The RNA transcript is then released to float around the cell. Some RNA will bind to a ribosome so it can then be translated into the end-product: a protein.

In the next chapter, Chapter 5, you will take your Genetic Engineering Hero journey further and discover the second of the *Three Steps to Microfacturing*, translation, by extracting some of the manufactured proteins you are engineering (microfacturing). You'll also learn more about how the cell takes RNA and converts it into amino acid chains called proteins.

Review Questions

Hands-on Exercise

1. Explain the difference between selective and non-selective plates

2. Explain negative and positive controls

3. What makes cells chemically competent?

4. Do you always recover your cells?

5. Why do you streak cells during a transformation experiment?

6. Why is it important to collect single colonies for a transformation?

7. Why is it important to grow streaked cells for 12-24 hours and no more?

Fundamentals

1. Describe the Four B's.

2. Do cells deliberately make decisions?

3. What are the *Three Steps to Microfacturing*?

4. What are the differences between RNA and DNA nucleotides? (Figure 4-19)

5. What is RNA polymerase?

6. What is needed to start transcription?

7. Which DNA strand does RNA polymerase "read"?

8. What are the three factors that can cause transcription to stop?

Chapter 5

Extracting your engineered proteins

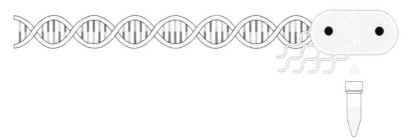

In Chapter 4 you made a monumental leap in officially becoming a genetic engineer! Congratulations!

By making LB agar plates, growing cells, making those cells chemically competent, then heat shocking, recovering and incubating them, you were able to insert DNA plasmids into some blank K12 *E. coli* cells. While your cells were recovering, the ones that took up a DNA plasmid immediately began the *Three Steps to Microfacturing*. We focused on the first step called transcription where DNA is read and transcribed by RNA Polymerase into a different nucleic acid, RNA.

The cells also simultaneously started the second of the *Three Steps to Microfacturing*, a process called translation. That's the primary focus of this chapter. Translation involves reading an RNA transcript and translating it into a chain of amino acids, called proteins. As mentioned in Chapter 3, proteins are usually the 'machinery' of cells, and in Chapter 4's hands-on exercise, you manufactured two different kinds of proteins: i) the protein color pigment; ii) the protein enzyme that enables the cells to be resistant to the antibiotic in the selective plates.

These are two examples of an unimaginable number of different proteins – each with a different function – being microfactured at any given moment across millions of species and organisms on planet earth.

The goal of this chapter's hands-on exercise is to combine and extend on the principles that you learned in Chapter 1 (lysing cells), Chapter 3 (growing cells), and Chapter 4 (engineering cells). This chapter will help you learn how to amplify and extract the proteins that you've engineered the cells to microfacture in order to start using them outside of the cells. Your final result will be a tube of 'cell extract' containing a large quantity of your microfactured product, such as your colored proteins.

Using your Minilab and microcentrifuge, you will:

- Grow and engineer cells with a DNA plasmid to make freshly engineered cells.
- Create several more selective LB agar plates.
- Streak the selective LB agar plates with your engineered cells to amplify them.
- Incubate your plates to allow for the growth of cells and expression of the proteins.
- Collect the cells and lyse them to extract your microfactured proteins.
- Centrifuge the samples to 'pellet' other 'cellular debris'.
- Sterilize the extract to remove any remaining bacteria.

In the *Fundamentals* section, we are going to dive into the second step of the *Three Steps to Microfacturing* - Translation. You will notice many similarities between how transcription and translation work. While different protein enzymes are involved, you will begin to see some general themes start to emerge. As these themes start to become clear in your mind, you will know that you are beginning to understand how cells operate and how to engineer them!

As with transcription in Chapter 4, we will cover: How cells know how to start translation, do translation, and stop translation. With a solid grasp of these fundamentals, you will know the basics behind the *Three Steps to Microfacturing*, and begin to understand the underlying chemistry behind genetic engineering. With this, you will be ready to take on the final step of microfacturing in Chapter 6, Enzymatic Processing, where you will also learn about atoms and bonding.

Getting Started
Equipment and Materials

The **Amino Labs' Engineer-it Kit™** and **Plate Extract-it Kit™** include all the required pre-measured ingredients. These kits be ordered separately at https://amino.bio/products

Shopping List

Amino Labs Engineer-it Kit™ (Optional) For this experiment, you can use your saved Engineer-it Kit S(e) or S(+) petri dish, refrigerated for no more than 1 week. Use a new Engineer-it Kit to get freshly engineered cells, or use the tube of pre-engineered cells that comes with the Plate Extract-it Kit™.

Amino Labs Plate Extract-it Kit™ (https://amino.bio)
Minilab (DNA Playground)
Microwave
Microcentrifuge

Instructions Overview

Prepare: Get Freshly Engineered cells
1. Use your freshly engineered cells from your refrigerated S(e) plate (if it is no more than 1 week old), repeat the Engineer-it Kit exercise from Chapter 4 to obtain fresh colonies of engineered bacteria, or use the pre-engineered cells that come in the Plate Extract-it Kit.

Day 1: Amplify your Engineered cells
2. Use the Plate Extract-it Kit to create selective LB agar plates. Pick fresh colonies to spread onto two selective LB agar plates.
3. Incubate the petri dishes for 24-48 hours until the streaked cells grow and express the desired trait (*e.g.,* color pigment).

Day 2: Extract proteins
4. Collect the cells by using an inoculating loop to scrape the cells from the surface of the selective LB agar petri dishes and deposit them into a tube of Lysis Buffer.
5. Lyse the cells by using Lysis Buffer and enzymes from the Extract-it Kit to break the cells open and release the microfactured products.
6. Incubate lysing cells at room temperature (color proteins) for 1 to 24 hours (or up to 72 hours if you incubate in the refrigerator).

Day 3: Sterilize proteins
7. Centrifuge the sample using a microcentrifuge to 'spin' the samples at max speed (13,000 x *g* or higher) to pellet the micelles and cell debris while leaving the extracted proteins dissolved in the liquid.
8. Filter sterilize the sample as there may still be viable bacteria in the extract.

Chapter Timeline Overview

Timeline to complete the hands-on exercise is:

Preparation: If you are completing a new Engineer-it Kit, this will add 3 or 4 days of hands-on exercise
Day 1: ~60 minutes followed by 24-48 hours incubation
Day 2: ~30 minutes followed by 1-72 hours incubation
Day 3: ~60 minutes

Timeline to read *Fundamentals* is typically 3 hours.

Learning Hands-On: Culture and lyse engineered *E. coli* to obtain a protein product extract

Step 1.   Download the instruction manual for the Plate Extract-it Kit

 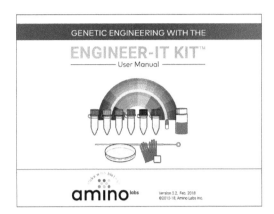

Figure 5-1. Step 1. Download the Plate Extract-it Kit instructions. You may also need the Engineer-it Kit Instructions.

Familiarize yourself with the Plate Extract-it Kit instructions found at https://amino.bio/instructions. These may reflect updates to the kit's material and instructions and also include what to look out for when completing the steps. If any conflict between these and the manufacturers' instructions happen, always follow the manufacturer's instructions.

You may also need the Engineer-it Kit instructions if you are doing the genetic engineering of bacteria again.

Step 2.   Put on your gloves and lab coat

Step 3.   Transform cells to get fresh colonies (Optional)

For this exercise it is best to have **fresh colonies of genetically engineered *E. coli* cells** on selective plates. If you have completed the exercise in Chapter 4 in the last week and your selective LB agar plates with engineered bacteria were stored in a way that they did not dry out (in a ziplock-style bag in a refrigerator), you can use a bacterial colony from one of those plates. If you have a new Engineer-it Kit, you can repeat the Engineer-it experiment. Practice makes perfect! By repeating the exercise, you will refine your transformation skills, which are amongst the most important for genetic engineers. Alternatively, you can use the pre-engineered cells that come with the Plate Extract-it Kit being aware that your best results would come from freshly engineered colonies.

When amplifying cells, it is always wise to use cells that have been freshly engineered. While this is not mandatory, it is good practice since the number of viable cells may be much lower on a plate that is not fresh. The longer you store your cells, the higher the chance that they have died. Further, the longer you store your plates, the higher the chance of seeing contamination. If you see anything that doesn't look like your engineered cells growing in your plates, this could be contamination and it is not recommended that you use any of the cells in that petri dish, even if they appear "clean" and separate from the contamination. If it is a fuzzy mold growing, spores could be throughout the plate and can contaminate plates that you're trying to culture and spread to other experiments!

Step 4.   Make selective LB agar plates for amplification

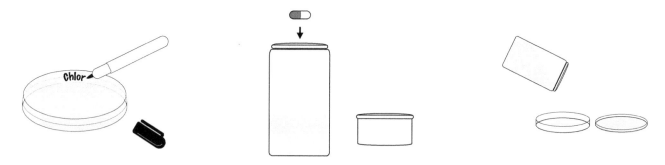

Figure 5-2. Step 4. Prepare your Selective Plates from the Extract-it Kit to amplify the quantity of cells for extraction.

Once you have your engineered cells, you must now amplify them. To amplify your cells, you will use **selective LB agar plates** made with your Plate Extract-it Kit. You will not need any non-selective plates for this experiment, so you can add the antibiotics immediately after heating and entirely dissolving your LB agar powder.

In the prior chapters, you labeled your petri dishes with an "S" for the antibiotic-containing selective plates. In this exercise, you can move one level closer to being a Genetic Engineering Hero by labeling the bottom your petri dishes with the antibiotic abbreviation instead of "S". Check within your Plate Extract-it Kit to see which antibiotic is used. In Chapter 3, you learned about labeling your plates with a colored streak to identify the antibiotic (Table 3-1). Another way to label plates is with an abbreviation for the antibiotics, as you saw in Table 4-1. For example, if you are using the antibiotic chloramphenicol, you would label your plate with the abbreviation "C" or "Chlor". As you become more sophisticated in your experimentation, it is common to use many different antibiotics to select for different engineered organisms. By labeling the specific antibiotic name or abbreviation on the bottom of the plate, you will make sure you are using the correct selective plates for the corresponding experiment.

Step 5.   Culturing: Spread out your freshly engineered cells

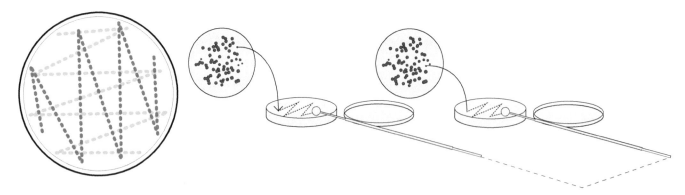

Figure 5-3. Step 5. Amplify your engineered cells by streaking them on two selective plates in a double zig zag pattern.

Culturing is a general term for amplifying your cells using petri dishes or in liquid broth culture. Here you will be spreading your engineered cells across **two of the four selective plates** to amplify them. The two remaining plates can be refrigerated for your own experimentation.

If you are using cells from the S(e) or S(+) plate, select a **fresh colony of engineered cells** that is expressing the deepest coloration and, using **a yellow inoculating loop** from the Plate Extract-it Kit, press the end of the loop into this single colony to transfer it to the loop so that you may spread it on your new selective plates. If you are using the tube of pre-engineered cells included in the Plate Extract-it Kit, dip your yellow inoculating loop in the tube to tranfer the cells, just like you did in Chapter 3.

Spread the cells onto the first selective LB agar plates by tracing a 'dual zig zag pattern' with your loop (Figure 5-3). Use the same loop for the entire spreading procedure as the goal of this exercise is to get lots of cells, like you did for your bacteria paint in Chapter 3. Repeat this again on a second selective LB agar plate using the same loop, but after selecting another colorful colony from the original plate, or dipping in the tube of pre-engineered cells again. At the end of this step, you should have two selective LB agar plates that have engineered bacteria spread across them.

Lysing cells: 'more doesn't mean better' *Pro-tip*

The Plate Extract-it Kit has enough lysis buffer to support two plates of cultured bacteria. If you add cells from more than two plates, you will not have enough surfactant to lyse all of the cells, and the experiment will not work as well.

This is an excellent example of the classic science rule; "more does not mean better". In fact, in scientific research and genetic engineering projects, you have to use optimal experimental conditions. If you were to use four plates of cultured cells and add them to the tube of lysis buffer from the Plate Extract-it Kit, you would get less 'cell lysis' than by adding just two plates of cultured cells. This is because there won't be enough surfactant molecules (Triton X-100) to cut through the membranes of the extra cells effectively. In trying to cut into twice as many cells, the molecules will be unable to lyse all the bacteria fully.

Imagine you have enough maple syrup to make 2 pancakes delicious. If you were to add 2 more pancakes to the stack without adding more syrup, the 4 pancakes would not taste that great because there is not enough maple syrup per pancake. Just as there is an optimal syrup to pancake ratio for each human, there is an optimal surfactant to cell ratio.

Step 6. Culturing: Incubate at 37 °C for 24-48 hours

Figure 5-4. Step 6. Incubate your petri dishes at 37 °C for 24-48 hours until you see your bacteria colonies grow and be colorful.

Just like in Chapters 3 and 4, it is recommended that you incubate the culturing cells at their optimal temperature, 37 °C for 24-48 hours.

Keep in mind!

The ideal temperature for *E. coli* growth is 37 °C. However, some proteins are more stable at lower temperatures so 30 °C or even room temperature can be used in some instances. If you like, you can incubate at 37 °C the entire time, at 30 °C the entire time, or a combination of both such as 37 °C for 24 hours and 30 °C for another 24 hours .

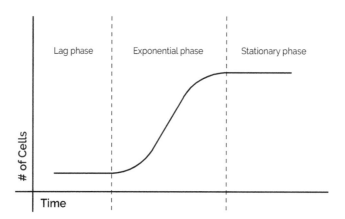

Figure 5-5. The S-curve describes the growth phases that organisms go through.

As you learned in Chapter 4, for transcription to start, sigma factors must bind to the promoter of a gene. This then enables the RNA polymerase to bind and begin transcribing the RNA from the DNA template strand. The sigma factor for the promoters controlling the transcription of Engineer-it Kit proteins is expressed in cells during a particular time of their growth. This stage of growth is called the "Stationary Phase" of cell growth, as seen in Figure 5-5.

The S-curve is a graph that is used to describe the number of *E. coli* cells (y-axis) as a function of time (x-axis). There are three major phases of *E. coli* growth:

1) Lag phase: This is the first phase where there are low numbers of cells, and the cells have abundant nutrients to support their growth. The cells begin growing rapidly.

2) Exponential phase: This is the phase when cell numbers increase the fastest. They have abundant food, space, and the temperature is optimal for metabolic processes. They have not begun sending each other chemical signals to slow down.

3) Stationary phase: This is the phase when nutrients become limited, and the cells begin sending chemical signals to other cells to slow their growth. During the stationary phase, cell division slows, and cells start preparing for the possibility of starvation.

It is in this third phase that your cells will really begin to start expressing the trait that you've engineered the cells to create - about 12-48 hours after you start incubating. The genetic engineer who designed these genes did this intentionally to make sure the cells would survive the transformation, start growing and then start to express the trait once the cells are abundant. If the genetic engineer had made the cells immediately express the trait, they might not have had enough energy and nutrients to survive. Asking the cells to create two other traits (antibiotic resistance and color proteins in this case) is quite taxing on their metabolism. Once the cells start slowing down in the stationary phase, they expend less energy on growing and can spare that to create the proteins and genes they have been engineered to express.

The cell metabolism has a delicate balance that genetic engineers must respect and experiment with when designing plasmids.

Step 7.   Extraction: Collect cells and start the lysis

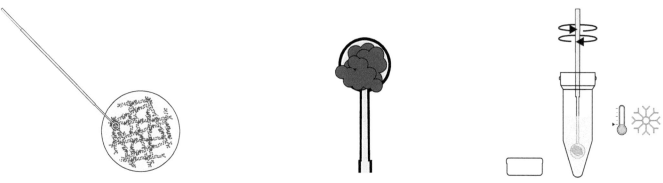

Figure 5-6. 7B. Scrape and mix your engineered cells in the Lysis Buffer.

After incubating your cells for 24-48 hours, a lot of cell colonies or lawns should have grown across the two petri dishes. The cells should now be expressing the desired trait, such as a color pigment.

Once they are producing their trait in a sufficient quantity (*e.g.* the color is very bright and saturated), collect the cells using a yellow inoculating loop and suspend them in the tube of Lysis Buffer. Use the cell-and-buffer mixing method you learned in the previous chapter. Continue collecting cells on the loop and mixing them into the Lysis Buffer until most of the cells on your plates are suspended in the Lysis Buffer tube:

A. Open your tube of **Lysis Buffer** and **Lysis Accelerator.** Place them in a tube rack or use the stations on your DNA Playground to hold them upright. Be careful not to spill!

B. Gently drag your **yellow inoculating loop** across the surface of the LB agar to collect the cells inside the loop. Once the loop has lots of cells, dip it into the Lysis Buffer tube and twist the inoculating loop like a blender to dislodge and mix-in the cells - just like you did when creating competent cells during the transformation process. Be careful not to be too vigorous; you don't want to splash liquid and cells around.

C. Repeat the scraping and blending process until you've collected the majority of cells from your two plates. Blend the cells and buffer for a further 60 seconds to make sure they are fully suspended. This will help the surfactant in the Lysis Buffer begin lysing the cells.

Lysis Buffer *Going Deeper* **5-2**

The Lysis Buffer contains a 'gentle surfactant' called Triton X-100. As in Chapter 1, when you lysed fruit cells with the surfactant SLS (from household soaps), here you will use Triton X-100. Triton X-100 is often used because while it breaks down cell membranes, it will not break down and destroy the proteins you've engineered the cells to produce. There are many different surfactants used for lysing cells, and Triton X-100 is one of the most widely used for extracting and isolating proteins.

Step 8.   Extraction: Lyse the cells

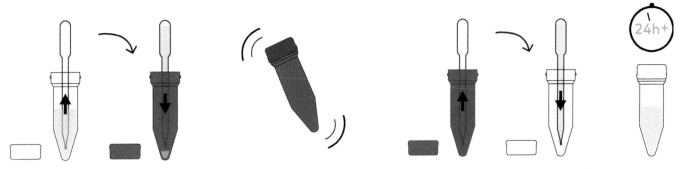

Figure 5-7. Step 8. Move your cells that are suspended in Lysis Buffer to the Lysis Accelerator tube, and back. Incubate.

Now that you've mixed your cells into the Lysis Buffer, they will begin to lyse. You will now introduce a new component to help aid in the lysis of the *E. coli* cells, the Lysis Accelerator.

The active ingredient in the Lysis Accelerator is a protein enzyme called Hen-Egg-White-Lysozyme (HEWL). HEWL is extracted and purified from the egg-whites of hens and has a unique function of being able to attack and break open bacterial membranes.

A. Using one of the **pipets** in the Plate Extract-it Kit, transfer all of the cells slurry from your Lysis Buffer tube into the brown Lysis Accelerator tube. You will be using this pipet again in a few minutes to transfer the slurry back into its original tube. Rest the pipet inside the Lysis Buffer tube for now.

B. Once you have moved all the buffer over to the Lysis Accelerator tube, firmly close the lid of the Lysis Accelerator tube. Then vigorously shake it for 30 seconds to ensure that it is thoroughly mixed. Tap the tube on the table to make all the liquid collect at the bottom of the tube.

C. Using the same pipet as in Step A, transfer your lysed cells into the Lysis Buffer tube. This will help you to complete the centrifugation Step 9, as it is easiest to do using a clear tube.

D. Let it stand at room temperature for 1 to 24 hours. You can also incubate them in the refrigerator if you want to incubate for 24 to 72 hours. You can shake the tube throughout this incubation to improve cell lysis.

HEWL *Going Deeper* 5-3

During this incubation, both the surfactant and HEWL are at work. Similar to the principles that you learned in Chapter 1, the surfactant is interacting with the outer and inner membranes of the cells, then binding up the lipids and some proteins into micelles. Since Triton X-100 is not as aggressive a surfactant as SLS, it is common to also use the enzyme HEWL to aid the process of cell lysis further.

Hen Egg White Lysozyme (HEWL), is a protein enzyme that is capable of binding to and breaking apart the peptidoglycan mesh in the intermembrane space of K12 *E. coli* cells. Recall from Chapter 3 that peptidoglycan is the amino acid-sugar hybrid that acts as a security layer in the intermembrane space of K12 *E. coli*. This is what the HEWL attacks and degrades. By using both HEWL and Triton X-100, you're able to gently break down the membranes of the cells so that the proteins you have microfactured can spill out.

Interestingly, humans produce enzymes that are similar to HEWL. These enzymes are produced in our tear ducts and skin and act as the first line of defense against bacteria in our eyes and on our bodies.

Step 9. Extraction: Pellet the cell debris

Figure 5-8. Step 9. Pelleted cells in a microcentrifuge tube.

Centrifuging enables you to separate materials of different densities. After the lysing process is complete, the cells, along with cell debris in micelles and other aggregates, will be denser (have more mass per unit volume) than smaller molecules that remain dissolved (like proteins and water molecules).

Microcentrifuges typically spin at 13,000 revolutions per minute (RPM) or more, which corresponds to an acceleration of more than 10,000 x g (italicized g is the force of gravity, not grams). This powerful force will cause the denser molecules to fall to the bottom of the tube in a compact mass called a "pellet". In this exercise, the pellet will include micelles, macromolecules like lipids, bits of cell membrane, and even genomic DNA. Small cellular components, such as proteins and DNA plasmids, may not be pelleted and can remain in the liquid.

Warning!

If this is your first time using a microcentrifuge, set it up as per the manufacturer's instructions. Then, start by spinning the microcentrifuge without tubes in it. Try different speeds, starting slowly. As the rotor speeds up, there should be no vibrations; it should be balanced. You should observe the same behavior when you add your balanced samples into the microcentrifuge; they should not cause the microcentrifuge to vibrate.

If there is any vibration, immediately stop the microcentrifuge and re-balance your samples.

A. Balancing the microcentrifuge. A microcentrifuge is meant for holding 'microcentrifuge tubes', which typically hold volumes between 0.5 mL and 2 mL. The Lysis Buffer, cells, and lysozyme in your tube of lysed cells will amount to about 1.1 mL, so your tube is probably quite full.

Your centrifuge should always be balanced. This means you should place another tube of equal mass in the open spot directly opposite from your Lysis Buffer tube. Inside the Plate Extract-it Kit, you will find a **Balancing Tube** which may already contain some water in it. If it does, note that while it will be close to the corresponding mass you want, you should always double-check to make sure it is correct. To balance your tube, use a **new clean pipet** to add tap water to the Balance Tube until the liquid level is similar. Use your small scale to weight the tubes and confirm they weight the same.

The tubes should be within 5% mass of each other. For example, if Tube A = 2 g, then Tube B should be no more than 2.1 g or less than 1.9 g because 5% = 0.05 x 2g = 0.1g. The closer the mass, the better. Make sure to close the tubes back tightly after balancing them.

B. With your experiment's lysis tube set opposite your balance tube, close the microcentrifuge lid(s) as per the manufacturer's instructions, tighten the lid, set the speed at the highest setting, and start the microcentrifuge. Monitor to see if the microcentrifuge vibrates - if it does, stop it immediately! Note that the microcentrifuge will hum and make some noises as part of its normal operation. A vibration means you either did not place your tubes opposite of each another, or you have not adequately balanced the tubes.

Spin at maximum speed (13,000 x g - 15,000 x g) for 20 minutes. During this time, your proteins stay dissolved in the liquid, while other macromolecules and cellular debris are forced to the bottom into a pellet.

At the end of this step, there should be a transparent liquid at the top that has your desired trait (*e.g.*, for a purple pigment, the liquid will be translucent purple) and, in the bottom of the tube, there should be a pellet of nearly white cellular debris. There may be some trait (*e.g.*, color pigment) remaining in the cellular debris and this simply means that not all of your cells lysed. If this occurs, in future experiments you can let your cells incubate longer in the Lysis Buffer and Lysis Accelerator.

Step 10. Extraction: Filter sterilize your proteins

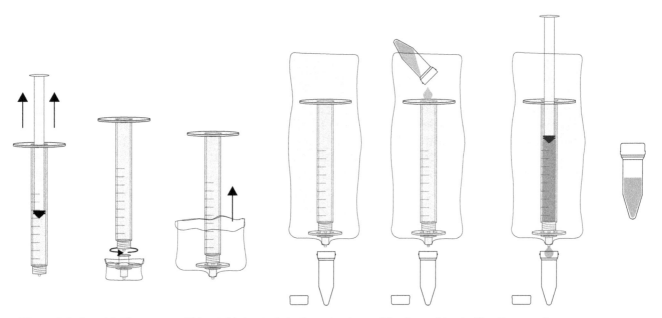

Figure 5-9. Step 10 -There may still be viable bacteria in the extract, so a filter is used to sterilize the sample.

Once you have lysed and pelleted the cell debris, there should be few remaining bacteria in the clear solution. We'll call this the sample. To be sure that there are no living bacteria or other microorganisms in the sample, you will now filter sterilize it. This not only helps ensure safety, but it also helps ensure that your samples will not spoil if stored for prolonged periods of time.

The filter provided in the Plate Extract-it Kit screws into the end of the supplied syringe to help press your sample through the filter. Within the filter, there is a membrane with very small pores that allow molecules like proteins, DNA and sugars through, but it traps larger molecules and entire bacteria. The filter provided has pores that are 0.22 um in size. Can you remember how large a typical *E. coli* bacteria is? You can find the answer in Chapter 3.

Caution!

In the next step, be aware that if you apply too much pressure to the syringe, the syringe may 'burst' or the filter could separate at the seam, resulting in a high-pressure stream of flying liquid. So do not press too hard while filtering your solution. It is recommended that you wear safety goggles and a lab coat.

If you find that it is getting increasingly harder for the liquid to pass through the filter, or has stopped passing through entirely, it is likely that cell debris from the pellet has clogged the filter. If this is the case, you should acquire a new filter.

A. Using your DNA Playground or a tube rack, set in the **Final Pigment Tube** and remove the lid.

B. Remove the syringe plunger from the **syringe** and lay it on a clean surface.

C. Open up the **syringe filter** packaging but DO NOT fully remove the filter from its packaging. Do this by either: a) taking the paper cover off if the filter is in a plastic/paper package, or, b) by opening the sealed plastic bag it is in. You do not remove the filter because you want to make sure that you do not contaminate it before use.

D. Holding the filter via the plastic container/bag, screw on the syringe to the filter, so it is firmly connected. You can lay this on the table, but be sure not to touch the sterile output end of the filter.

E. Extra Protection: Use the **Burst Bag** in the Plate Extract-it Kit to enclose the syringe and filter. The Burst bag is not a foolproof solution, but in the rare event that a burst event happens, the bag will act as the first line of blockage for a spraying sample.

F. After prepping your Burst Bag, and placing your syringe/filter inside it, gently pour or pipet your Lysis sample into the open syringe. Do this gently to avoid causing the pellet of cell debris from falling into the syringe. If this happens, the pellet will clog up the syringe. If you see clumps of cell debris pour into the syringe, pour all the contents back into the tube and repeat the centrifugation step.

G. With the sample in the syringe, hold it so that the sterile end of the filter points into the Final Pigment/ Product Tube. Replace the syringe plunger into the syringe and GENTLY but FIRMLY press down. If you have effectively microcentrifuged your sample, the plunger will slowly push in until all the solution passes through. Cell debris or bacteria in the sample will be trapped in the filter. Small molecules like your proteins will pass through. In the event that most but not all liquid passes through before cloggging the filter, this tells you that next time, you should incubate in the lysis tube longer or centrifuge a bit longer.

H. Close and tighten the lid on your Final Product Tube. Congratulations! You have now sterilized the proteins you microfactured using your genetic engineering skills! You can store your final pigment in the refrigerator, or at room temperature. Many color pigments will keep their color for more than a year if kept out of the sun.

Step 11.   Using your proteins

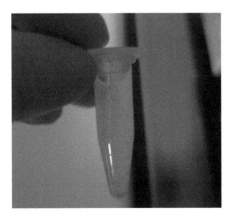 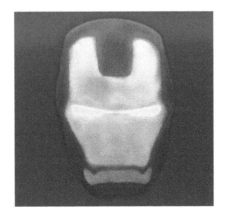 

Figure 5-10. Examples of fluorescent pigments extracted from colored proteins engineered in *E. coli* with Amino Labs Engineer-it Kit and Extract-it Kit: i) Fluorescent proteins under UV light; ii) Etched plastic dyed with fluorescent Yellow, Cyan, and Magenta proteins by J. Pahara, 2017; iii) Frog painting made with Fluorescent Yellow and Cyan extracted proteins by John from Toronto, Canada, 2016

This cell extract with a high concentration of the colored protein can be used in different ways (Figure 5-10):

- If it is a fluorescent protein, place it next to a black light to see it glow

- Express yourself artistically! Try using it as ink or watercolor. You can use different types of paper and drawing instruments like paintbrushes. If you have a fluorescing protein and plan to use a black light to illuminate your artwork, note that a black light can cause some light papers to glow blue, which may affect your artwork; place a black light next to the paper to test before.

- Wear your sample proudly, or even gift it: Find a small container that can hold liquid. Seal the opening and attach the container to a necklace, pin or other wearable item!

- Try dyeing fabrics or other materials with it! You can even layer it onto etched or porous plastic. See if you can dye that surface!

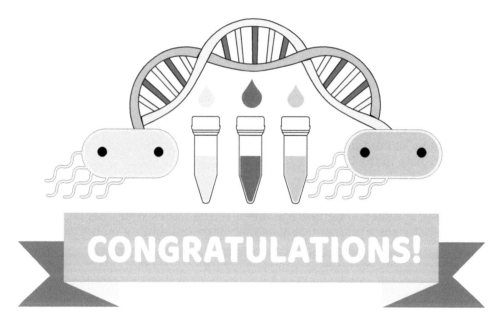

Fundamentals: How cells translate proteins from RNA

Step two of the *Three Steps to Microfacturing*: Translating proteins from RNA

Let's pick up from where we left off in Chapter 4. In your hands-on experiment, when you inserted DNA plasmids into cells, RNA transcripts of genes were transcribed from DNA templates by RNA polymerases. Depending on the sigma factor and how strongly it bound to the promoter, there could be tens, hundreds, or thousands of RNA transcripts from your gene of interest bumping around the K12 *E. coli* cell. Each transcript becomes a key part of the second step of Microfacturing - translation.

Just like how RNA polymerase "reads" DNA, binds ribonucleotides and strings them together to form a strand of RNA, translation involves a different cellular machine called a ribosome that binds to and "reads" RNA. While reading the RNA, it binds up amino acids and strings them together to form proteins. If you've forgotten what amino acids are, check back to Chapter 3. Also, just like we saw in Chapter 4 how DNA has coding and non-coding regions, RNA also has coding and non-coding regions (Table 5-1). These regions determine when and how much protein is translated and where to start translating from.

The equivalent of a promoter in DNA is the ribosomal binding site (RBS) in RNA. In other words, in a DNA molecule, the promoter helps the transcription machinery know where to start transcription, and in an RNA molecule, the RBS helps the translation machinery to know where to start translation.

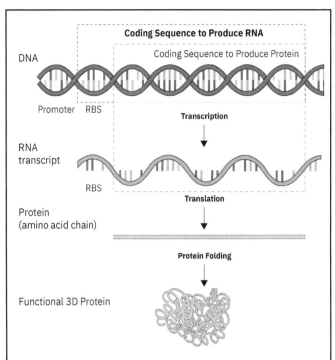

Figure 5-11. During the first two steps of the *Three Steps to Microfacturing*, DNA and RNA each contain coding and non-coding regions. Upon the creation of the protein chain during transcription, it folds into a 3D structure called a protein. It is the 3D structure of the protein that determines its function.

In Figure 5-11, notice that DNA contains all the information for transcription and translation: Promoter, RBS, and the coding sequence. As mentioned in Chapter 1, DNA is the master blueprint of the cell. During transcription, embedded in the DNA coding region is an RBS - the RBS does not have any function during transcription. Only once the RNA transcript is created, will the RBS, which is now at the 5' phosphate end of the RNA transcript, will become relevant and be

Nucleic Acid	Non-coding region	Position	Function
Table 5-1. Non-coding regions that function in DNA and RNA			
DNA	Promoter	The promoter is upstream of the RNA coding region and determines the DNA strand that the RNA polymerase transcribes.	Binds to sigma factors which then can bind to and orient RNA polymerase to commence the creation of RNA transcripts through the process of **transcription**
RNA	Ribosomal binding site (RBS)	In DNA, the RBS is just downstream of the promoter but is still upstream of the protein coding region. In RNA the RBS is at the 5'P end of the transcript.	Binds to initiation factors which then can bind to the ribosome and can commence the creation of proteins through the process of **translation**

key in causing translation to start. The region downstream of the RBS is the RNA coding region, which codes for the protein to be created by the translation machinery.

During translation, there is the creation of the amino acid chain (protein). The folding of that chain into a three-dimensional shape results in the protein structure that has a function.

Starting Translation

How does the cell machinery know when and how to start translating an RNA transcript? Translation involves similar principles as in transcription: Proteins are bumping around the cell and can bind to the ribosomal binding site in RNA, which can then also bind to a ribosome - these protein are called initiation factors.

Proteins already made by the cell called initiation factor 1 (IF1), initiation factor 2 (IF2), and initiation factor 3 (IF3) bump around in the cell. While these are three different proteins, they all work together to start the process of translation. In accordance with the Four B's, the initiation factors bump around until they interact with the ribosome. The ribosome's purpose is to read the RNA transcript sequence and translate the RNA sequence into an amino acid chain. An amino acid chain is commonly referred to as a protein. Just like RNA polymerase, the ribosome will use a cipher to do this, which we will explore in the next section. The initiation factors also help to start translation by binding to the RNA transcript. If the ribosomal binding site (RBS) has the right shape and charge to bind to one or more initiation factors, the initiation factor(s) will bind.

For the translation process to be successful, when the ribosome binds to the RNA via the initiation factors, it must also lock into the RNA so that it does not "fall off". This is similar to how, during transcription, the RNA polymerase created a short piece of RNA called the initiation sequence. However, the ribosome is slightly different thanks to a special "built-in" feature that pre-equips it with the locking mechanism. As you can see in Figure 5-12, the ribosome is made up of both protein and RNA intertwined with one another. That's right, the ribosome is actually a hybrid of both protein and RNA! It is the **ribosomal RNA** that allows it to lock into the RNA strand through complementary ribonucleotide interactions (A-U, G-C) (Figure 5-12, Figure 5-13).

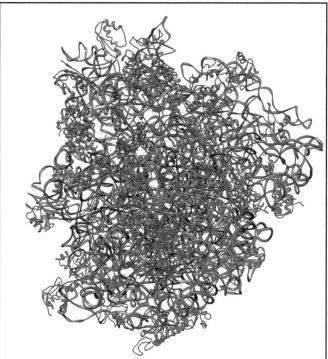

Figure 5-12. A crystal structure of a ribosome. Ribosomes bind to and read RNA and translate their sequence into a sequence of amino acids - also called a protein. The ribosome itself is a mixture of both protein and RNA that function in harmony. Purple: "16s RNA" strand; Red: other RNA strands; Blue: Protein. Crystal structure data from A. Korostelev, S. Trakhanov, M. Laurberg and H.F. Noller (2006) Crystal Structure of a 70S Ribosome-tRNA Complex Reveals Functional Interactions and Rearrangements. Cell 126:1065-1077.

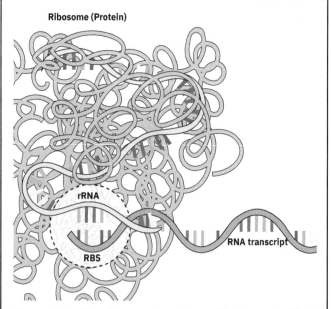

Figure 5-13. The ribosome is made up of both protein (blue) and RNA (yellow). It is the ribosomal RNA (rRNA, yellow) that allows the ribosome to lock into the RNA transcript (orange) by complementing the ribosomal binding site (RBS) at the 5'P end of the RNA transcript (orange).

This ribosomal RNA (rRNA) is known as 16s rRNA and is intertwined within the protein structure of the ribosome. 16s rRNA is normal RNA, except it doesn't get translated into a protein, and simply stays as RNA. The rRNA folds upon itself in a way that allows it to also merge with the ribosome protein (Figure 5-12). The rRNA is an essential part of the ribosome because it is what enables the protein structure to lock into the RNA transcript thanks to complementary binding of ribonucleotides. While the initiation factors help the ribosome bind to the RNA initially, it is the 16s rRNA that gets the ribosome in position and readies it for translation. Just like two complementary DNA strands can come together, the rRNA complements a short sequence of the ribosomal binding site of the RNA transcript (Figure 5-13).

During Translation: The RNA to protein cipher

The initiation factors do the Four B's and bind to the ribosome, which further bumps around and becomes bound to the RNA transcript at the RBS. The rRNA in the ribosome complements and bonds to the RBS of the RNA transcript. Now that the ribosome is locked into the RNA at the RBS using the rRNA, it can start the process of translation. Translation involves "reading" the RNA transcript while simultaneously creating an amino acid string.

Let's briefly revisit how RNA polymerase works during transcription, as it has some simimlarities to how the ribosome works in translation. The RNA polymerase uses a cipher to "read" DNA and transcribe RNA (Table 4-2). The RNA polymerase cipher is based on the complementarity of the DNA and the ribonucleotides (A's bind to U's, C's bind to G's, T's bind to A's). While the RNA polymerase "reads" the DNA, millions of ribonucleotide molecules (A's, U's, G's, C's) bump around and, when the "right" ribonucleotide "fit in" to the RNA polymerase and complements the DNA nucleotide being read by the polymerase, the RNA polymerase permanently attaches it to the growing chain (Figure 4-29).

Translation also has a cipher that relies on complmentarity and the Four B's, but is slightly more complicated than the one for transcription. Let's explore it now, along with the machinery that the ribosome uses to "read" RNA and create a chain of amino acids. Unlike in transcription, where the RNA polymerase simply "adds on" ribonucleotides that complement the nucleotides in the DNA template strand, amino acids cannot

directly complement the RNA transcript. Therefore, the ribosome needs a "go-between" to bridge the gap between the RNA transcript and the amino acid. These "go-betweens" are another kind of hybrid molecule called transfer RNA (tRNA) (Figure 5-14). tRNA is a hybrid molecule made up mostly of RNA and one amino acid: the one end of the tRNA is able to interact with and complement the RNA transcript through what is called an anticodon, and on the other end is an amino acid that the ribosome can add onto the growing amino acid chain. Coincidentally, the tRNA also sort of has a hand-written cursive capital "T" shape (Figure 5-14).

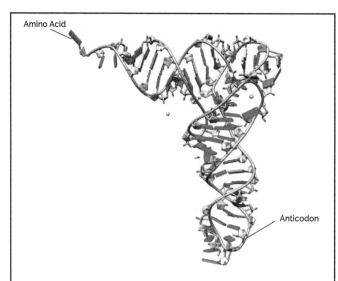

Figure 5-14. A transfer RNA (tRNA) molecule is a string of non-coding RNA that folds into a "T" shape and has an amino acid at one end while the other end binds to the RNA transcript. It folds into this shape simply because of complementary regions of RNA nucleotides. See the hairpin structure of an RNA terminator in Chapter 4.

Quite a lot happens in translation in order for the RNA to be translated into a protein. Let's pause and summarize all of the different players involved in translation:

- **RNA transcript/messenger RNA:** The RNA polymerase, a protein enzyme, transcribed the information from DNA to make the RNA transcript during transcription. Another name for the RNA transcript is messenger RNA (mRNA). The RNA transcript has a non-coding region called a ribosomal binding site (RBS), as well as a coding region which is what will ultimately be translated into a protein.

- **Initiation factors:** Initiation factors bind to the ribosome and RBS of the RNA transcript. They are analogous to the sigma factors in transcription.

- **Ribosome**: The ribosome is a hybrid of protein and RNA, and it catalyzes the chemical reactions for joining amino acids together. The ribosome is analogous to the RNA polymerase in transcription.

- **rRNA**: Ribosomal RNA (rRNA) is embedded in the ribosome protein and helps it to lock onto the RNA transcript's ribosomal binding site (RBS) just prior to when translation starts. It does this by complementing (zippers with) the ribonucleotides that make up the RBS.

- **tRNA:** Transfer RNA (tRNA) are specialized "go-betweens" that complement the RNA transcript coding region and have an amino acid bound to its other end. They bridge the gap between RNA and amino acids.

  A tRNA with an amino acid attached to it is called an **amino**acyl-**tRNA** (Figure 5-15), but they are commonly referred to just as tRNAs.

During translation, the ribosome "reads" three ribonucleotides from the RNA transcript at one time using tRNA. It does this because each tRNA anticodon complements three ribonucleotides in the RNA transcript coding region at once. A group of three ribonucleotides in an RNA transcript are called a **codon** (Figure 5-15). In other words, the three letter codon of the RNA sequence (*e.g.*, AUG) complements the anticodon sequence of a tRNA (*e.g.*, UAC). Similar

to how during transcription nucleotides bump around the cell until the right one bumps its way into the RNA polymerase and becomes attached to the growing transcript, tRNAs bump around the cell and when the right tRNA bumps into a ribosome that is loaded with mRNA and the anticodon complements all three ribonucleotides in the codon, the ribosome adds the amino acid on that tRNA to a growing chain of amino acids (Figure 5-16 and Figure 5-17). In other words, it takes three RNA nucleotides of information to encode one amino acid.

Each amino acid is encoded by one or more RNA codons (Table 5-2). You'll see in Table 5-2 that there are 64 possible RNA codons. Proteins are typically chains of 50 amino acids or more which require 50 codons or more. This means that the RNA transcripts needs to be at least 150 ribonucleotides, plus the non-coding RBS and terminator. This further means that the DNA gene must be 150 nucleotides, plus promoter (for transcription), the RBS (for translation), and the terminator (for transcription)! The RNA codon table (Table 5-2) is used to know which RNA codons code for each amino acid. Note that different codons can code for the same amino acid. For example, C-A-U and C-A-C both encode for the amino acid histidine. This means that there is an aminoacyl-tRNA (tRNA) that, at one end, has an RNA anticodon sequence that complements the C-A-C or C-A-U, and has a histidine amino acid at the other end.

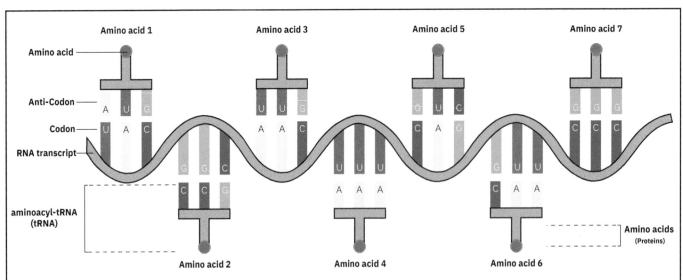

Figure 5-15. tRNAs are the physical connectors that bind to the RNA codons in the ribosome. When the appropriate tRNA binds to a codon (and they complement), the amino acid at the "base of the T" is added to the growing amino acid chain.

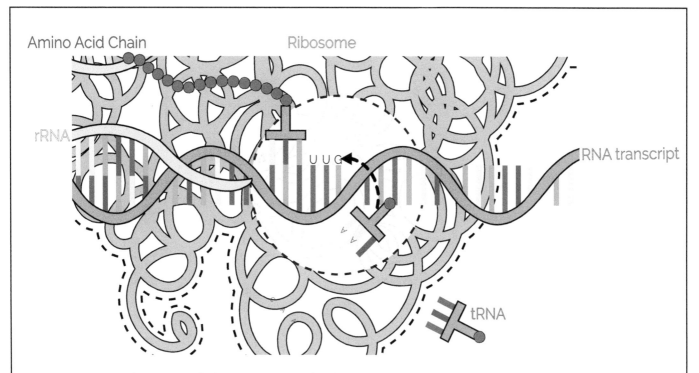

Figure 5-16. As the ribosome travels downstream, it "reads" the RNA transcript. A bumping tRNA that is complementary to the RNA codon in the ribosome will bind strongly, and this triggers the ribosome to attach the amino acid end to the growing amino acid peptide strand. If the three ribonucleotides (codon) and tRNA anticodon do not bind strongly, the tRNA will bump out of the ribosome, and eventually the correct tRNA will bump into the ribosome and be added to the growing string of amino acids.

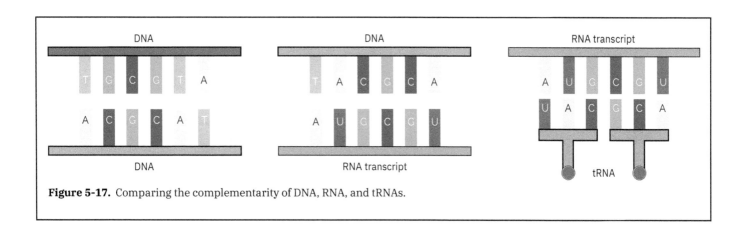

Figure 5-17. Comparing the complementarity of DNA, RNA, and tRNAs.

tRNA *Going Deeper* 5-4

In Figure 5-15, you can see the tRNAs as "T" shaped, with an amino acid at the bottom and three ribonucleotides at the top. While it is true that in their real physical form the tRNAs are somewhat T shaped (Figure 5-14), the tRNAs of Figure 5-15 are a cartoon representation of the 3:1 ratio, and not faithful to that physical shape. Visit the Protein Data Bank webpage to see more in-depth information about tRNAs:

https://pdb101.rcsb.org/motm/15

Table 5-2. Translation cipher - RNA codon table

1st base	2nd base								3rd base
	U		C		A		G		
U	UUU	F/Phe Phenylalanine	UCU	S/Ser Serine	UAU	Y/Tyr Tyrosine	UGU	C/Cys Cysteine	U
	UUC		UCC		UAC		UGC		C
	UUA	L/Leu Leucine	UCA		UAA	Stop (Ochre)	UGA	Stop (Opal)	A
	UUG		UCG		UAG	Stop (Amber)	UGG	W/Trp Trypto-phan	G
C	CUU		CCU	P/Pro Proline	CAU	H/His Histidine	CGU	R/Arg Arginine	U
	CUC		CCC		CAC		CGC		C
	CUA		CCA		CAA	G/Gln Glutamine	CGA		A
	CUG		CCG		CAG		CGG		G
A	AUU	I/Ile Isoleucine	ACU	T/Thr Theronine	AAU	N/Asn Asparagine	AGU	S/Ser Serine	U
	AUC		ACC		AAC		AGC		C
	AUA		ACA		AAA	K/Lys Lysine	AGA	R/Arg Arginine	A
	AUG	M/Met/Methi-onine	ACG		AAG		AGG		G
G	GUU	V/Val Valine	GCU	A/Ala Alanine	GAU	D/Asp Aspartic acid (Aspartate)	GGU	G/Gly Glycine	U
	GUC		GCC		GAC		GGC		C
	GUA		GCA		GAA	E/Glu Glutamic acid (Glutamate)	GGA		A
	GUG		GCG		GAG		GGG		G

Four B's for Choosing Correct Amino Acid: *Going Deeper* 5-5

It should be emphasized that similar to RNA polymerase in transcription, the ribosome does not have the "intelligence" to read, analyse, and understand the RNA sequence and then use that information to decide which amino acid to attach to the growing chain. Rather the Four B's are at the centre of it all! Recall the Four B's concept you read about at the beginning of the *Fundamentals* in Chapter 4: Bump, Bind, Burst (optional), and Bump.

In the case of the ribosome, there are many tRNAs with amino acids attached to them that are produced by the cell and are bumping around inside the cell. This pool or tRNAs is also capable of bumping into and out of the ribosome. As a ribosome is positioned at a particular spot on the RNA, three ribonucleotides are available to bind to the correct tRNA. tRNA bump in and out of the ribosome, and when the correct tRNA that has an anti-codon that matches and binds to the RNA codon, this triggers the ribosome to complete the next step of attaching the amino acid to the growing amino acid chain (protein).

So while the ribosome doesn't have human-like intelligence and decision making skills, the "information" that determines which aminoacyl-tRNA is correctly chosen is embedded within the system in the form hydrogen bonding between the ribonucleotides of the codon and anti-codon! In other words, the information is "hardwired into the system".

Be the cell Machinery! DNA to Protein *Breakout Exercise*

In this super quick breakout exercise, you're going to practice what you know about how a cell is able to **"read"** DNA and ultimately create protein.

Table 5-3. Practice the transcription and translation ciphers				
DNA + leading strand	AGG	GAG	GCC	GAT
DNA- template strand				
RNA (codon)				
tRNA (anti-codon)				
Amino acid (protein)				

*Remember that in RNA there are no "T's"! Refer back to Table 4-2

Protein Data Bank (PDB) of Ribosome *Web Search Breakout*

The Protein Data Bank (PDB) is an online database of X-ray crystal structures of biomolecules. There is a fantastic overview of ribosomes and how they work. https://pdb101.rcsb.org/motm/121

During Translation: Locating the starting point for translation

How does the ribosome know precisely where to start translating the coding region of the RNA transcript? You've already learned that the initiation factors 1, 2, and 3 are responsible for recruiting the ribosome and the RBS at the 5' end of the RNA transcript, and the rRNA in the ribosome lock it into place. Let's now learn how, once the ribosome is positioned and ready to go, it knows exactly where to start.

While it may seem like the RBS would be the starting point for translation, the RBS is actually an approximation of the start point and only serves to lock in the ribosome close to the start point. This is similar to typing - when you start typing, your hands will hover over the keyboard, with your fingers near, but not on the keys you will soon be using - you are locked into your keyboard but not yet using it. In the RNA transcript, the starting point is specific: a single codon from the 64 possible RNA codons is called Start Codon, and it is required to be near the RBS for

the ribosome to start translating. The start codon is also the same codon that encodes for the amino acid methionine. Have a look in Table 5-2, what is the three-letter RNA codon for Methionine/Met?

A-U-G! The start codon will be found 2 to 6 ribonucleotides downstream of the ribosomal binding site (Figure 5-18). Once the ribosome binds to the ribosomal binding site, the initiation factors, which are also bound to a special "starter" methionine tRNA called fMet (formylmethionine), help kick-start the translation process. The initiation factors "give" the fMet to the ribosome to become the first amino acid of the protein and then the initiation factors leave. Only for the A-U-G start codon is the fMet used and any A-U-G codon that is present inside of a sequence being translated, uses the regular methionine tRNA. This is in part because once translation starts, the initiation factors leave the ribosome and because the initiation factors have the unique ability to bind the fMet and give it to the ribosome during translation initiation, fMet doesn't usually play a role during the rest of translation.

After the start codon, the ribosome continues to move along the coding region of the RNA transcript (Figure 5-18). Thousands of the 64 different tRNA molecules complete the Four B's of basic cell operation and, when the correct tRNA enters a translating ribosome and its anticodon matches perfectly with the RNA transcript codon present in the ribosome, the ribosome takes and connects the amino acid to the growing chain (Figure 5-16). As new tRNA come into the ribosome, it releases the previous tRNA that lost its amino acid to the growing protein chain.

As the ribosome is translating the amino acid chain, the amino acids in the chain begin to interact with one another. The 20 amino acids are each quite unique - some are negatively charged, some positively charged, and some not charged (Figure 3-30). Some of these have sizeable bulky side chains (arginine), and some have small side chains (glycine). It is this wide variety of amino acid side chains that gives the amino acids different "personalities". As the amino acid chain gets longer during translation, the different amino acids in the chain bump and bind or repel each other to form a complex and beautiful three-dimensional shape (Figure 3-28). This three-dimensional shape of proteins is what results in their function - like being colorful!

In Figure 5-19, you will find one product of translation, the crystal structure of a protein called Red Fluorescent Protein (RFP). Just like the structure of DNA you saw in Chapter 1 (Figure 1-19), you will see the red, blue, and gray segments of this structure

which correspond to oxygen, nitrogen, and carbon, respectively.

In Figure 5-19 you will also see a "ribbon" throughout the structure, and this is commonly used as a guide to help visualize the intricate and beautiful "beta barrel" structure of the protein but is not a physical part of the protein. RFP is only one example of thousands of proteins that *E. coli* make, all of which have different 3D shapes. You'll see that RFP is very small at 3.5 nanometers wide. Compared to a DNA helix at ~12 nanometers wide, proteins can be quite small.

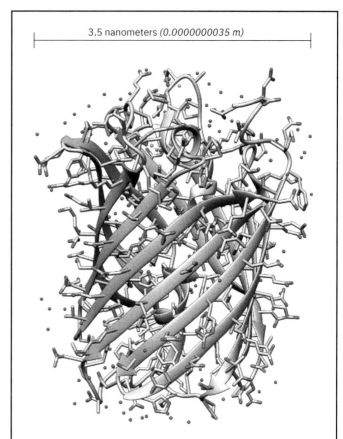

Figure 5-19. A protein is a long chain of amino acids folded up into a three-dimensional structure. In this example, red fluorescent protein (RFP) has a beta-barrel structure. RFP is a small protein that is about 250 amino acids long and is about 3.5 nanometers wide. Crystal structure data source: RCSB PDB: 2VAD.

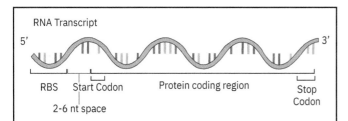

Figure 5-18. RNA transcripts are transcribed by RNA polymerase and are made up of ribonucleotides. A transcript has a ribosomal binding site (RBS), 2-6 nucleotide spacer, start codon, protein coding region, and a stop codon.

The methionine start codon *Going Deeper* **5-6**

So does this mean that every protein starts with methionine? Yes, it does, at least when it is made initially. *E. coli* cells have some protein enzymes called proteases that are able to clip off the ends of proteins after translation. Typically, proteases will cut off several amino acids from either end of the protein, one of which could be the starter methionine.

Stopping Translation

How does the ribosome know when to stop translating the RNA transcript? It is quite simple; at the end of the coding region, there is a stop codon (Figure 5-18).

If you look back at Table 5-2, you'll notice that three of the sixty-four codons are marked as "Stop". These are codons that do not bind to tRNA molecules but instead bind to release factors. Release factors are proteins that are able to recognize the specific codons in the RNA transcript causing the amino acid chain to become disconnected and "fall out" of the ribosome. When this occurs translation stops!

Choosing codons when designing DNA *Pro-tip*

If you're designing DNA for your own biotechnology project, you might be asking the question: Which codons do I use? There are often several codons for the same amino acid, so how do I know which one is best. Do I get to choose whichever I want?

The short answer is that you base your decision on which host organism you plan to express your protein in. For example, if you choose K12 *E. coli* bacteria, like the ones you have been using in your hands-on experiments, you will choose codons based on which codons K12 *E. coli* have evolved to use.

As you learned in Chapter 4 and this chapter, the end goal of expressing a gene is not always to get proteins. Rather, there are genes that express extremely important RNA molecules that have no intention of being read by a ribosome - tRNA is a perfect example.

tRNAs are not expressed equally in cells, rather, there is a diversity of tRNA expression within a single organism, and between species of organisms. For example, if we were to extract and purify all the tRNAs listed in Table 5-2, we would not see them in equal proportions. Rather, *E. coli* cells express some tRNAs more than others. See *E. coli* Codon Use Table below (Table 5-3).

In Table 5-3 you will see the codons separated by the amino acid they code for in alphabetical order. If you look at "A", which stands for alanine, you see four different codons, all of which will cause the ribosome to add an alanine to the amino acid chain. You'll notice that of the four codons, GCG is used the most (33% of the time), whereas GCC, GCA, and GCU are used less so at 26%, 23%, and 18%, respectively. As a general rule, the codon use generally reflects the amount of tRNAs that are expressed in the cell. In the case of alanine tRNAs, the CGC (anti-codon) tRNA will most highly expressed, while the CGA (anti-codon) tRNA will be least expressed.

Based on this knowledge you can design your DNA so that the overall codon use in your gene are consistent with the codon table. If you don't follow this general design rule, and say you only use the GCU codon (18%), there may be too few alanine tRNAs available during translation of your proteins. Recall back in Chapter 4: The number of molecules available for a chemical reaction is an important part of the Four B's and can determine whether a reaction will or will not occur. The low number of available CGA tRNAs means the Four B's cannot do the job and the ribosome can simply stop translation.

Designing DNA to have the right codons seems like a lot of work, right!? Nowadays you can find "codon optimization" software on the web, and/or when you order DNA from a DNA synthesis company, they will automatically optimize the DNA sequence based on your target host.

As a final note, you might now be asking, why is Table 5-3 a codon table and not an anti-codon table? The answer to this will lead you down a rabbit hole of information that goes beyond the scope of this book. If you'd like to follow this thread, search for "all anti-codons in *E. coli*", and/or "wobble base pairs".

Codon	Amino acid	Codon fraction for same amino acid	Fraction of total codon use (%)*	Codon	Amino acid	Codon fraction for same amino acid	Fraction of total codon use (%)*
UAA	STOP	61%	0.2%	AUG	M	100.00%	2.64%
UAG	STOP	9%	0.03%	AAU	N	49%	2.06%
UGA	STOP	30%	0.1%	AAC	N	51%	2.14%
GCU	A	18%	1.71%	CCU	P	18%	0.75%
GCC	A	26%	2.42%	CCC	P	13%	0.54%
GCA	A	23%	2.12%	CCA	P	20%	0.86%
GCG	A	33%	3.01%	CCG	P	49%	2.09%
UGU	C	46%	0.52%	CAA	Q	34%	1.46%
UGC	C	54%	0.61%	CAG	Q	66%	2.84%
GAU	D	63%	3.27%	CGU	R	36%	2%
GAC	D	37%	1.92%	CGC	R	36%	1.97%
GAA	E	68%	3.91%	CGA	R	7%	0.38%
GAG	E	32%	1.87%	CGG	R	11%	0.59%
UUU	F	58%	2.21%	AGA	R	7%	0.36%
UUC	F	42%	1.6%	AGG	R	4%	0.21%
GGU	G	35%	2.55%	AGU	S	16%	0.99%
GGC	G	37%	2.71%	AGC	S	25%	1.52%
GGA	G	13%	0.95%	UCU	S	17%	1.04%
GGG	G	15%	1.13%	UCC	S	15%	0.91%
CAU	H	57%	1.25%	UCA	S	14%	0.89%
CAC	H	43%	0.93%	UCG	S	14%	0.85%
AUU	I	49%	2.98%	ACU	T	19%	1.03%
AUC	I	39%	2.37%	ACC	T	40%	2.2%
AUA	I	11%	0.68%	ACA	T	17%	0.93%
AAA	K	74%	3.53%	ACG	T	25%	1.37%
AAG	K	26%	1.24%	GUU	V	28%	1.98%
CUU	L	12%	1.19%	GUC	V	20.00%	1.43%
CUC	L	10.00%	1.02%	GUA	V	17%	1.16%
CUA	L	4%	0.42%	GUG	V	35%	2.44%
CUG	L	47%	4.84%	UGG	W	100%	1.39%
UUA	L	14%	1.43%	UAU	Y	59%	1.75%
UUG	L	13%	1.3%	UAC	Y	41%	1.22%

Translation is more precise than transcription: During translation, there is a very specific start point (start codon) and a very specific stop point (stop codon) in every RNA sequence. Remember that transcription, on the other hand, is much more 'sloppy'. The promoter region in DNA facilitates binding of RNA polymerase. As long as the RNA polymerase can create the initiation sequence and escape the promoter, it begins transcribing 'roughly' where it binds at the promoter. Transcription completes in a sloppy manner by rho-dependent factors chasing RNA polymerase and knocking it off, or by rho-independent factors such as a 'hairpin' and/or a poly-U repeat region causing the RNA polymerase to fall off the DNA.

Why do genetic engineers program in DNA when ultimately the cells read RNA to make their desired proteins? While RNA is the programming language cells actively use to make proteins, it is neither very stable nor durable. DNA, however, is excellent for long-term information storage as it is very stable. Remember that one of the main differences between DNA and RNA is the sugar present in the molecules; while RNA is made up of ribose sugar, DNA is made up of deoxyribose. Deoxyribose is simply ribose with one less OH group. This lack of OH aids the stability and durability of the DNA because that extra OH group makes the RNA molecule more susceptible to hydrolysis. Hydrolysis is a process during which a molecule breaks down from a reaction with water. While this does make RNA a less stable "programming language" for Genetic Engineers, some have started investigating and using RNA hydrolysis as a feature in their design.

Be the cell machinery! Decode the message *Breakout Exercise*

Your turn! Decode the secret message from DNA with the two ciphers you learned in Chapter 4 and 5.

Table 5-4. Transcription cipher - Nucleotide pairing table		
DNA + Leading Strand	**DNA - Template Strand**	**RNA**
A	T	A
T	A	U
C	G	C
G	C	G

Table 5-5. Be the RNA Polymerase & the Ribosome															
	5'													3'	
Leading DNA (+)	gat	gaa	tgc	att	ccg	cat	gaa	cgc	gat	aac	gcg	ccc	gcg	ggc	gag
Template DNA (-)															
RNA															
Protein															3	6

Be the cell machinery! Bidirectional translation *Breakout Exercise*

Now that you have learned more about how the ribosome translates an RNA transcript into protein using the RNA to protein cipher, head back to page 120 (Ch. 4) to finish the bidirectional translation Breakout Exercise.

Similar to how you found that the RNA transcripts had different sequences when transcribed from the DNA strand in opposite directions, you'll also find that the protein sequences are different. Other details to note:

- the presence of starter methionines
- stop codons
- recall where translation starts (*e.g.* do you translate the RBS?)

While in this exercise your proteins are only five amino acids long (called a peptide), in many real genetic engineering scenarios, your DNA sequence would be hundreds or thousands of deoxyribonucleotides long, leading to an RNA transcript that is hundreds or thousands of ribonucleotides long, and ultimately an amino acid sequence (protein), that is hundreds to thousands of amino acids long. For example, the colourful proteins that you engineered your K12 *E. coli* cells to produce in Chapter 4, and then extracted in this chapter:

- have a DNA sequence, including promoter, that is ~1000 deoxyribonucleotides (or basepairs/ bps) long
- have an RNA sequence, including the RBS, that is ~800 ribonucleotides long
- have an amino acid sequence that is ~250 amino acids long

Summary and What's Next?

Congratulations! In this chapter, you not only genetically engineered the microorganism *E. coli*, cultured it in a selective LB agar plate, lysed the cells, extracted and sterilized the protein that you engineered *E. coli* to microfacture, you also took your understanding of the *Three steps of microfacturing* further! This is a massive accomplishment! These are foundational skills and knowledge that every genetic engineer needs to know to engineer and manipulate cells.

You can see that from a DNA molecule with a promoter, RNA is transcribed with the help of sigma factors and RNA polymerase. If that RNA transcript molecule has an RBS, initiation factors in the cell can interact with the transcript and a ribosome. The ribosome uses its rRNA to lock into the RBS and commences translation with the help of tRNAs and the initiation factors holding an fMet. tRNAs complement the codon triplets in the RNA transcript to a specific amino acid. As the ribosome moves downstream on the RNA transcript, amino acids are bound to a growing chain of amino

acids. Once the ribosome hits the stop codon, the peptide chain is released, and it can finish folding into a three-dimensional shape using chemical bonding, a topic of Chapter 6. This is how and why there are now many beautiful three-dimensional proteins floating in the cells you engineered.

In Chapter 6, you'll go through Step 3 of the *Three Steps of Microfacturing: Enzyme Processing*. Enzyme processing is not always necessary. In many cases, the product of translation is a protein that is itself the desired product. For example, you've created a color protein pigment. The function of that protein is to be colorful, and that's it! Microfacturing stops here. In many other instances, however, the protein created from translation is an enzyme that is meant to be used to cause chemical reactions to happen. In Chapter 6, you're going to learn how you can engineer cells to create an enzyme that can catalyze chemical reactions you can then use to your benefit.

There's more to translation... *Web Search Breakout*

The mechanism that the ribosome completes during translation is slightly more complex than described in this chapter. If you're keen to learn the full story, search the web for videos about this subject. Search "RNA translation" or "EPA sites ribosome".

Review Questions

Hands-on Exercise

1. Why is it important to have fresh colonies for culturing?

2. Why should you label plates?

3. How is the double streak method different from normal streaking?

4. What is the active ingredient in the Lysis Accelerator? Where does it come from? How does it help during cell lysis?

5. What is a pellet?

6. Why should you filter sterilize your extracted sample?

7. What is "balancing a centrifuge" and why is it important?

Fundamentals

1. What is the equivalent of a promoter in RNA?

2. Describe why a ribosome is a hybrid structure

3. What is the difference between RNA transcript and rRNA?

4. What is the difference between the codon and the anticodon?

5. How does the process of translation use the Four B's?

6. What is fMet?

7. What are the three different "kinds" of RNA you have learned about? Explain the differences.

Check Point!

You've accomplished quite a lot on your journey so far, congratulations! Below is a 19 point checklist which summarizes what we learned and did in Chapters 1-5. Review it before going further, and make sure to repeat any exercises or breakouts if you feel it necessary!

☐ LB agar plates are made. They are the food (sugar, amino acids) and substrate (solid agar scaffolding in a petri dish) that *E. coli* can grow on. Non-selective LB agar plates are used to grow "Blank cells", while selective LB agar plates are used to grow genetically engineered cells that have a selection marker.

☐ Blank cells are collected from a stab, streaked onto a non-selective LB agar plate and incubated at 37 °C for 12-24 hours so that fresh, fast-growing, non-engineered individual colonies can be used for the genetic engineering experiment.

☐ Chemically competent cells are made in a cold environment by collecting fast-growing *E. coli* cells and mixing them into Transformation Buffer. Transformation Buffer is a liquid that contains salts such as calcium chloride so that the cells are better able to take in DNA.

☐ A DNA plasmid is mixed into the chemically competent cells, incubated and then heat shocked at 42 °C to help the DNA plasmids pass through the cell membrane and cross into the cells. The cells are cooled to trap the DNA inside.

☐ The cells are recovered in Recovery Media at 37 °C to begin their growth cycle, division, and so that they start expressing their selection marker, such as an antibiotic resistance enzyme.

    ☐ To express the selection marker, the DNA plasmid includes a specific gene. This gene sequence has a promoter, RBS, and protein coding sequence (with a start and stop codon). After transcription and translation, the gene results in the expression of a protein enzyme called chloramphenicol acetyltransferase that can break down the antibiotic:

        ☐ Sigma factors bind to the promoter and recruit the RNA polymerase which creates an initiation sequence. It then tries to escape the promoter

        ☐ If the RNA polymerase escapes the promoter, it moves downstream on the DNA molecule, unzipping it and, as the correct complementary ribonucleotide enters the polymerase, adds it to the string of RNA

        ☐ Rho-dependent (Rho proteins) or rho-independent factors (poly-Ts and/or terminator hairpins) cause the RNA polymerase to "slip off" the DNA. The RNA polymerase then releases the RNA transcript into the cytoplasm of the *E. coli* cell - transcription ends.

        ☐ Initiation factors bumping around the cell bump into the RBS sequence of the RNA transcript and the ribosome which also binds to the RBS and locks in thanks to ribosomal RNA (rRNA).

        ☐ If a start codon is present, a special starter tRNA (fMet) which is bound to the initiation factors, binds to the ribosome and kicks starts translation. The ribosome rides down the RNA transcript, and many aminoacyl-tRNAs that are complementary to the particular RNA codon in the RNA transcript allow the ribosome to know what amino acids to add to the growing amino acid string.

        ☐ The ribosome encounters a Stop Codon where a release factor resides and concludes the translation of the amino acid string. The ribosome falls off the RNA transcript

        ☐ While translation is happening, as well as afterward, the amino acid string folds upon itself to form a three-dimensional macromolecule called a protein. It is the three-dimensional structure of the protein that determines its function.

☐ After 24-72 hours, colonies of bacteria can be seen on the selective experimental LB agar plates. The colonies on the plate will be expressing the trait that they were engineered to show, such as a change in color due to the creation of a colored protein.

☐ Just like a sample of bacteria was collected from a stab, some bacteria from colonies on the plate were collected and used to culture, or amplify bacteria on further selective LB agar plates.

☐ Cells are incubated for 24-48 hours until they sufficiently expressed their trait (such as color). During this time, the cells were completing transcription and translation, which resulted in the trait.

☐ Cells packed full of your colored proteins are collected and added to a Lysis Buffer which contained a surfactant called Triton X-100, and a Lysis Accelerator called HEWL which was an enzyme that actively degraded the peptidoglycan in the intermembrane space. The samples are incubated so that the cells lysed, and the colored proteins were released.

☐ The lysed cells are microcentrifuged at high speeds (more than 10,000 x g) so that all the cell debris is pelleted at the bottom. The sample's supernatant is poured into a syringe with a microfilter attached.

☐ The sample is sterilized by passing it through a 0.22 um pore diameter filter and stored in a tube for later use.

Chapter 6

Processing Enzymes

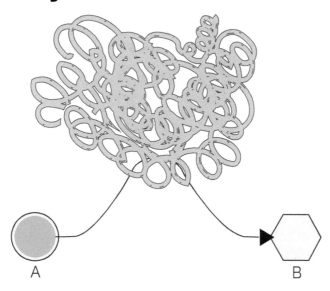

In Chapters 4 and 5 you genetically engineered cells with DNA plasmids that contained two genes, one that coded for antibiotic resistance, and another to express a trait such as a chromoprotein (color pigment). The proteins that resulted from the K12 *E. coli* cells transcribing and translating the genes had very different functions:

1. **Chromoprotein (color pigment):** This protein's primary function is to be colorful. Once the protein is created, it bumps around the cell and gives the cell color. The color pigment DNA was initially derived from a sea creature which benefits in some way from being colorful.

- **Antibiotic resistance:** This protein's function is to catalyze a chemical reaction. The protein created for chloramphenicol antibiotic resistance is called chloramphenicol acetyltransferase. The enzyme is able to bind to the chloramphenicol molecule and cause a chemical reaction that inactivates chloramphenicol and prevents it from being able to harm the cell.

This chapter focuses on the latter, protein enzymes that catalyze chemical reactions and how they are used by scientists and genetic engineers to create end-products. Steps 1 and 2 of the *Three Steps to Microfacturing* are used to create proteins from a DNA sequence (Figure 6-1), and, in many instances, the

protein itself is the end-product. However, in many other instances, the protein is not the end-product, it is instead used to catalyze a chemical reaction involving other molecules which lead to the desired end product(s). This is what we consider Step 3 of the *Three Steps to Microfacturing*: the processing of molecules using protein enzymes.

To explore the third step of microfacturing, you will see first hand in the hands-on exercises how enzymes can be used to convert one molecule into another. Using cell extract, you are going to make pure oxygen using a naturally occurring protein enzyme called catalase. You are going to complete a set of new genetic engineering experiments whereby you engineer K12 *E. coli* cells to express protein enzymes that will help you generate smells and a new set of colors in a test tube!

In the *Fundamentals* section, we are going to dig deeper into atoms, bonding, how enzymes function and the underlying chemistry of enzyme processing. We will then relate this topic back to cell metabolism and how it relates to "Life". You will then fully realize that the trillions of cells that make up you (and *E. coli*) are packed full of enzymes, all of which are encoded by genes stored in your cells' genome. These enzymes drive your metabolism, the chemical reactions that keep you going.

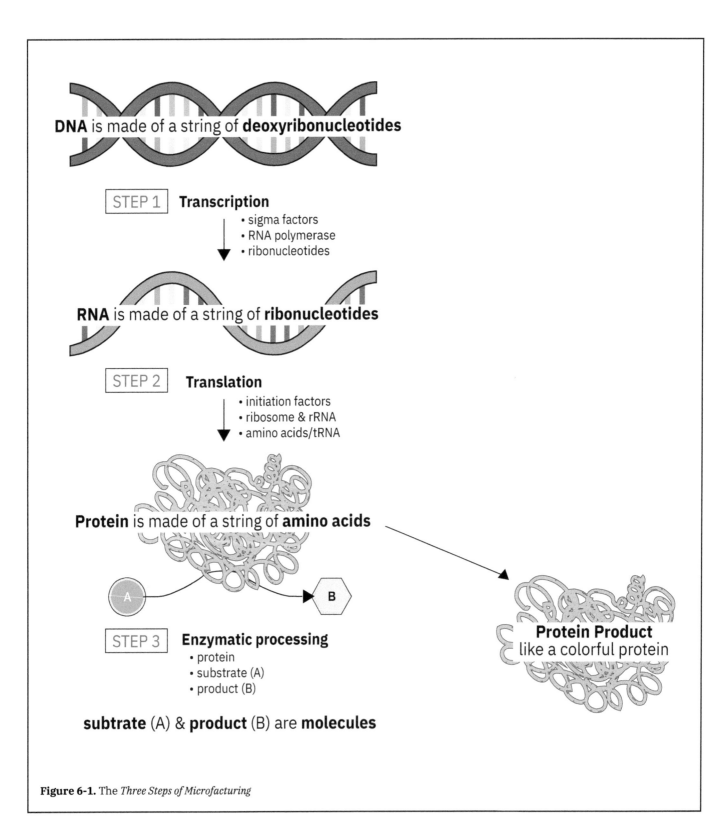

DNA is made of a string of **deoxyribonucleotides**

STEP 1 **Transcription**
- sigma factors
- RNA polymerase
- ribonucleotides

RNA is made of a string of **ribonucleotides**

STEP 2 **Translation**
- initiation factors
- ribosome & rRNA
- amino acids/tRNA

Protein is made of a string of **amino acids**

A → B

STEP 3 **Enzymatic processing**
- protein
- substrate (A)
- product (B)

subtrate (A) & **product** (B) are **molecules**

Protein Product
like a colorful protein

Figure 6-1. The *Three Steps of Microfacturing*

Getting Started
Equipment and Materials

The **Amino Labs' Smell-it Kit**™ and **Blue-it Kit**™ are contained in the *Zero to Genetic Engineering Hero Kit Pack Ch. 5-7* and include all the required pre-measured ingredients. These kits can also be ordered separately at https://amino.bio/products

Shopping List

Exercise 1:
Wetware kit: Amino Labs Smell-it Kit™ (https://amino.bio)

Exercise 2:
Wetware kit: Amino Labs Blue-it Kit™ (https://amino.bio)
Microcentrifuge
Small scale

Exercise 1 & 2:
Minilab (DNA Playground)
Microwave

Post-practice exercise:
Standard hydrogen peroxide (available at pharmacy)
Cell extract from the Blue-it Kit

Instructions Overview (for both Exercises)

Day 1-4 - "Bag 1"
1. Complete the genetic engineering exercises in "Bag 1"

Day 4-6/7 - "Bag 2"
2. Complete the culturing exercises using "Bag 2" to obtain cultured cells or cellular extract containing the engineered protein enzyme.
3. Add substrate to the experiment to complete Enzymatic Processing.

Chapter Timeline Overview

Timeline to complete each hands-on exercise is:

Day 1: ~60 minutes followed by 12-24 hours incubation,
Day 2: ~60 minutes followed by 12-24 hours recovery
Day 3: ~30 minutes followed by 24-48 hours incubation
Day 4: ~45 minutes followed by 24-48 hours incubation
Day 5: ~45 minutes followed by 24-72 hours incubation
Day 6: ~45 minutes
Day 7: ~45 minutes (Blue-it kit only)

Timeline to read *Fundamentals* is typically 3 hours.

Learning Hands-On: Process one molecule into another using enzymes

You've learned the basics of how to grow, engineer, culture, and lyse cells to obtain a cell extract that contains your microfactured proteins. We are now going to put those skills to work! In these hands-on exercises, you are either going to culture engineered cells and process molecules within the petri dish, or you are going to use your cell extract to convert one type of molecule, called a substrate, into a different molecule, called a product (Figure 6-1).

Chapter 6 includes two different exercises that demonstrate different ways to do enzymatic processing. Exercise 1 involves engineering bacteria to create an enzyme called Atf1 that converts a substrate into another with an overripe banana smell. This will be done within cells in the petri dish. Exercise 2 includes engineering cells to create an enzyme called beta-galactosidase that converts a colorless substrate into a blue product. This will be done in using cell extract.

Exercise 1: Enzymatic processing to generate smells

Step 1.   Download the instruction manual for the Smell-it Kit

Familiarize yourself with the Smell-it Kit instructions at https://amino.bio/instructions. The manufacturer's instructions will have the most up to date procedures for this exercise.

Warning: This experiment includes odors! If you are sensitive to odors or perfumes (give you headaches, make you nauseous, etc.), the odors in this kit could trigger your sensitivity.

Step 2.   Complete the engineering part of the Smell-it Kit

Complete the Smell-it Kit **"Bag 1"** genetic engineering portion of this exercise just as you have in Chapter 4. Whereas in prior genetic engineering experiments your primary objective was to engineer cells to produce color pigments, in this exercise you will engineer the cells to produce a protein enzyme. Upon successfully engineering the cells, they will appear visually similar to the blank cells that you streak on the non-selective LB agar plates. This is because the enzyme does not have a color. Do not be fooled! If they grow on selective plates, the plasmid is being expressed.

Step 3.   Culture cells with the substrate

Within the Smell-it Kit **"Bag 2"** you will find the culturing materials and tubes of substrate. The substrate is called **isoamyl alcohol** and has a musky smell. Open up the tube and gently **waft** your hand over the top toward your nose to try to smell the liquid. (Look at the *Smelling reagents Pro-tip* coming up if you are not familiar with wafting). Put the lid back on until you complete the next steps.

Smelling reagents *Pro-tip*

When smelling materials in any lab or engineering environment, you should never place your nose directly over the top of the container opening and inhale. This is very dangerous because the material you want to smell could be toxic or overwhelming. Wafting is a safe technique to use when smelling reagents.
Visit: https://amino.bio/wafting

Step 4. Labeling and creating your LB agar plates for culturing

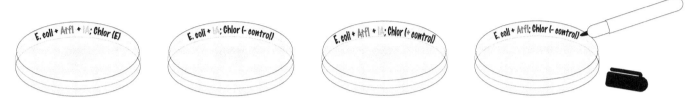

Figure 6-2. Label your petri dishes for your experimental sample and three controls

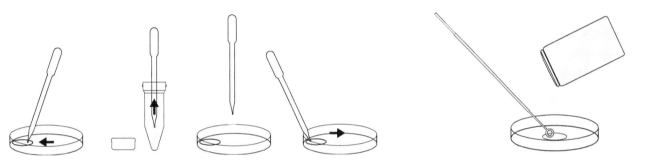

Figure 6-3. Transfer IA into petri dishes, and slide the paper disk over top of the IA to trap it underneath.

Figure 6-4. Pour molten LB agar into the petri dishes. Use a yellow loop to hold the paper disks so they won't float away.

A. You will use **four selective LB agar plates** for the culturing part of the Smell-it Kit experiment. Label them:

- ***E. coli* + Atf1 + IA; Chlor (E):** This is your experimental sample. If everything is done correctly, you should smell overripe banana. The organism is *E. coli* expressing Atf1. The plate will contain the substrate isoamyl alcohol (IA). The antibiotic is chloramphenicol.

- ***E. coli* + IA; Chlor (- control):** This is a negative control. The organism is negative control *E. coli* cells and does not express Atf1. The plate contains the IA substrate. The antibiotic is chloramphenicol. These cells will grow on selective LB agar plates, but will not express the Atf1 protein and therefore won't process the substrate. Because there is substrate, the plate should smell similar to the original substrate in the tube.

- ***E. coli* + Atf1 + IA; Chlor (+ control):** This positive control includes you streaking a plate with positive (+) cells included in the kit that express Atf1. You will add IA to the plate. Similar to the other plates chloramphenicol is the antibiotic for selection.

- ***E. coli* + Atf1; Chlor (- control):** This is a second negative control. In this instance, you will streak your plate with some Atf1 expressing engineered *E. coli* (+ cells). However, you do not add IA substrate. This control will verify that the *E. coli* themselves do not produce the overripe banana smell.

B. Place your **IA substrate tubes** in your centrifuge and "pulse" spin them down for 10 seconds. This will ensure that all the substrate is at the bottom - This is how Genetic Engineering Heroes guarantee all of their samples are at the bottom of the tube.

C. Create your selective LB agar as you have previously, but DO NOT POUR IT YET! Bring your **sterile distilled water** to a rolling boil. Add your **LB agar powder** and swirl to dissolve. Heat it further in 4-second intervals until you see the molten LB agar foam (boil). Add your **antibiotic** and swirl to dissolve it as you have done in prior exercises. DO NOT POUR!

D. While your LB agar is cooling, you must complete this step rather quickly. Using the **pipet** included in your kit, push the paper disks to the side of the three petri dishes. Using the pipet, transfer an entire tube of IA substrate to the middle of one of the plates and then using the pipet push the paper disk over the top to trap the IA underneath. Complete the same action with the other disks, tubes of IA substrate, and plates.

E. Pour your selective LB agar into each plate. Use a **sterile yellow loop** to hold the paper disk on the bottom of the IA dishes when you pour in the LB agar - the disks may begin to float, but you can keep them on the bottom of the plate by holding them with a yellow loop. You may use the same yellow loop for all plates. Allow your dishes to cool and solidify. The paper disk will help to release the IA substrate into the LB agar slowly. This slow release will create a sustained overripe banana smell that will last between 1-3 days.

Step 5.   Culture engineered cells on your selective LB agar plate

If your DNA Playground incubator has 8 plate capacity, you can streak and incubate all four plates at the same time. If you have a two plate incubator, you can streak/incubate two of four plates in any order - the cells will grow within 24 hours, after which you can streak and incubate the next two plates.

Following the plate labels you previously made, use the dual streak method (Chapter 5) for culturing cells to ensure you have a lot of cells growing across the surface of the plate. As the IA substrate diffuses to the surface, the cells will absorb the substrate, and the expressed enzymes will process isoamyl alcohol into a different molecule called isoamyl acetate - which has the smell of overripe (sweet) bananas.

As you open the lids of the plates, the smell will disappear and you will need to wait for it to replenish. Try to do a blind smell test by enlisting someone's help!

ATF1 Banana Smell *Going Deeper* **6-1**

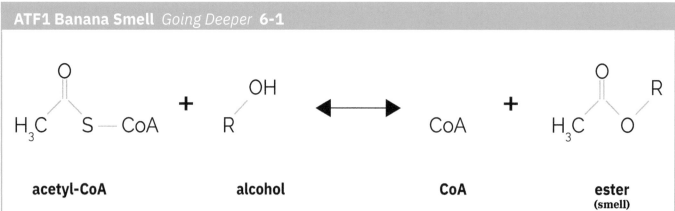

Figure 6-5. Atf1 enzyme takes an acetyl group from acetyl-CoA and transfers it to the isoamyl alcohol.

The protein enzyme that you have engineered the K12 *E. coli* to produce is called alcohol acetyltransferase 1, abbreviated Atf1. Atf1 catalyzes a chemical reaction whereby the musky smelling substrate, isoamyl alcohol, is converted to overripe banana smelling isoamyl acetate (Figure 6-5). When you added the isoamyl alcohol substrate underneath the LB agar, it slowly diffused into the LB agar, all the way to the surface where your engineered cells are. The substrate then crossed the outer membrane, intermembrane space, and inner membrane of the cells into the cytoplasm where many Atf1 enzymes were expressed. The chemical reaction includes three essential players:

- isoamyl alcohol substrate
- acetyl-CoA substrate (already created by the cells and is residing in the cytoplasm)
- Atf1 enzyme that catalyzes the reaction

Atf1, isoamyl alcohol, and acetyl-CoA complete the Four B's around the cell. When both substrates bind to the Atf1 enzyme, a chemical reaction occurs. A video describing how both substrates are changed by the chemical reaction can be found below. In short, a small chemical acetate group is transferred from the acetyl-CoA to the isoamyl alcohol. The products of the reaction include a CoA and the isoamyl acetate ester (Figure 6-5). You can find an explanation of the Atf1 chemical reaction at https://amino.bio/atf1

By adding the small acetate group on the isoamyl alcohol, you substantially change the characteristics of the molecule - it now makes a noticeably different smell!

Usually, this reaction doesn't happen without ATF1, so did you notice a difference between your ATF1+IA plates compared to your negative controls? Also, have a look back to Going Deeper 3-7 to refresh your memory about how enzymes help to make chemical reactions happen.

Exercise 2: Enzymatic processing to generate color

In Exercise 1, you engineered K12 *E. coli* cells to create Atf1, a protein enzyme that is capable of modifying alcohols into esters. This changes the chemical structure and also the physical odor properties of the molecule.

In Exercise 2, you will complete another enzymatic processing experiment, except this time using the cell extract rather than in a petri dish. In this experiment, you will be engineering *E. coli* cells to express an enzyme called beta-galactosidase. This enzyme is useful for processing a certain substrate molecule that does not have any color into colorful molecules. The substrate in this reaction is typically called X-gal, where the "X" can be one of many colors, and the "gal" standing for galacto. Two X-gal molecules are included in the Blue-It Kit: Blue-gal with its chemical name 5-Bromo-4-chloro-3-indolyl-β-D-galactopyranoside, and Yellow-gal with its chemical name 4-nitrophenyl β-D-galactopyranoside. Visit amino.bio to see if new colors are available!

In the reaction, the beta-galactosidase enzyme binds a water molecule to X-gal and removes a sugar from the X-gal molecule. When enzymes use water to complete a chemical reaction, it is called hydrolysis. Two of the hydrolyzed X-gal molecules then dimerize (come together in a pair) resulting in the molecules' physical properties changing and becoming colorful (Figure 6-6).

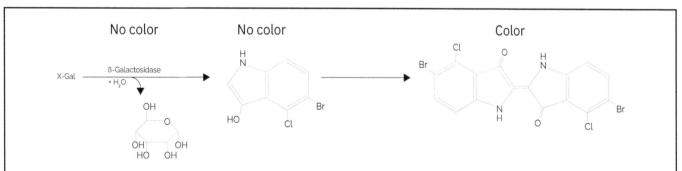

Figure 6-6. Beta-galactosidase removes a sugar ring from X-gal with the help of a water molecule. The products then "dimerize" into a colorful molecule.

Step 1.   Download the instruction manual for the Blue-it Kit

Find the manual at https://amino.bio/instructions.

Step 2.   Complete genetic engineering and extraction procedures

Similar to Chapters 4 and 5, complete the genetic engineering using the Blue-it Kit to obtain colonies of E. coli genetically engineered to express the beta-galactosidase enzyme. Using the cultured genetically engineered bacteria, complete the extraction to obtain a tube of sterilized cell extract containing the beta-galactosidase enzyme that you engineered the cells to express. A new and important aspect of this extraction is that the beta-galactosidase is sensitive to oxygen when outside of the cells and so you need to add a special additive to your Lysis Buffer. Called a "reducer", dithiothreitol (DTT) is a small molecule that has a sulfur group that makes DTT able to combat the harm that oxygen can have on beta-galactosidase and keeps beta-galactosidase from breaking apart. In your kit, you'll create a tube of DTT, called Enzyme Stabilizer, by combining the DTT powder in the Enzyme Stabilizer tube to the Enzyme Stabilizer Dissolving Buffer. You will use the Enzyme Stabilizer twice, once in this step and once when catalyzing your reaction. Make sure you put your tube of Enzyme Stabilizer in the freezer after this step.

You'll definitely notice that DTT is smelly and this is due to the sulfur atom that is part of the molecule! As Genetic Engineering Hero's do, spin your DTT powder tube in the Enzyme Stabilizer tube in a microcentrifuge to ensure that all of the DTT powder is at the bottom of the tube. Next, add the Enzyme Stabilizer Dissolving Buffer using a small pipet. Pipet up and down a few times to mix, until the powder is dissolved.

As per the instruction manual, add your Enzyme Stabilizer to the Lysis Buffer. Then, scrape and add your cells to the Lysis Buffer tube. Add the Lysis Accelerator as you have in the past. Incubate as recommended in the instruction manual.

After incubating your lysis reaction, microcentrifuge your sample and complete the filter-sterilization as you did in Chapter 5. You should now have a tube of sterile beta-galactosidase enzyme extract, also called your cell extract, for use in your experiment. Note that you will also need some of this cell extract for the post-practice exercise where you make pure oxygen from hydrogen peroxide. You will not be using all of it for the Blue-it Kit reaction so store the tube of cell extract in the refrigerator after using it in the next step. Further note that you'll want to complete the post-practice exercise within one day of refrigerating your cell extract.

Step 3.   Dissolve your substrates

Within the Blue-it Kit, there are two tubes labeled **"Blue-gal Substrate"** and **"Yellow-gal Substrate"**. These tubes have a small amount of white powder in them, and you first need to dissolve the powders with **Reaction Buffer**. Reaction Buffer is mostly water but has the right pH (acidity) to make sure the enzymatic processing works well, and the enzyme stays stable.

Similar to Step 2, you'll need to add Enzyme Stabilizer to each tube of Reaction Buffer.

The Blue-gal and Yellow-gal are shipped to you in powder form because they are not very stable when dissolved in a liquid. This is why you will dissolve them in the Reaction Buffer just before adding in your cell lysate. Similarly, the DTT is not very stable in other liquids, and that is why you dissolve it just before you complete the experiment. Refer to the instruction manual to complete this step.

Step 4.   Add cell extract beta-galactosidase to the substrate

Get ready for chemical wizardry! Using a **pipet**, add a small amount (50-100 uL) of cell extract to each substrate tube. The plastic pipet will have a small 100 uL marker on it to help guide you. Place the lids on firmly and invert them to mix them thoroughly. Place the tubes on your DNA Playground Hot Station set to 37 °C and sit back and let the beta-galactosidase do its job! Over the course of the next 24 hours, the samples will become increasingly blue and yellow. You'll see yellow start to appear in minutes (usually within 10 minutes), and the blue will appear over the course of hours.

Enzymatic Processing *Going Deeper* **6-2**

You have now completed enzymatic processing! In both exercises, you engineered the cells to produce protein enzymes (Atf1 & beta-galactosidase) that were able to complete the Four B's and ultimately interact with substrate molecules (isoamyl alcohol + acetyl-CoA or Xgal+H_2O) and turn them into products. But why did you do the Smell-it Kit within the cells *vs.* the Blue-it Kit as an extract? These are both useful methods and choosing between them depends on the many factors such as the stability of the protein enzymes, the availability of the substrates in the cell, and the cost of buying substrates.

In Exercise 1, a key reason for using streaked cells on plates, rather than an extract, was that one of the two substrates, acetyl-CoA, is very expensive to purchase and you needed the cell to keep microfacturing it naturally for your chemical reaction. Because of this, lysing the cells would have stopped the cells from making more acetyl-CoA, and you would have had only a small amount of acetyl-CoA for the reaction (when you lyse the cells, they will not create any more acetyl-CoA). Doing the reaction in living cells means the cells continue to grow and they can keep producing acetyl-CoA, and so long as you have lots of IA available, the reaction can continue to occur. This means you could keep adding IA to the cells and they would keep processing it into isoamyl acetate! For reference, 0.1 grams of acetyl-CoA costs more than $1000 to buy! Including it in the kit would sure increase the cost of the Smell-It Kit! Let the cells make it for you!

In Exercise 2, only one substrate molecule is needed for the chemical reaction, and that chemical is reasonably priced. There is no need to have the cells continue growing to create a substrate. Because a single beta-galactosidase enzyme can cause the chemical reaction to happen thousands to millions of times during its lifetime, you can simply use the extract with the engineered enzyme and add the X-gal substrate.

An alternative protocol for Exercise 2 could involve growing cells in a petri dish, similar to Exercise 1. Pouring or pipetting a small amount of the X-gal substrate over the top the plate of cells, or adding it into the LB agar when you pour the plates is common. The substrates in the LB agar can cross into the cells. The enzymes inside the cells will begin catalyzing the same chemical reaction inside of the cells - the colonies of bacteria would turn from white to colored.

Naming Enzymes *Pro-tip*

Protein enzymes are usually easy to spot because their name will end in "...ase". Generally, the beginning prefix of the enzyme name will give you some historical or chemical context about what the enzyme does, and then "ase" suffix is added to the end.

In Exercise 1, alcohol acetyltransferase (Atf1) is a great example. One might deduce from its name that it binds alcohols and transfers an acetyl group to it. In Exercise 2, the enzyme is beta-galactosidase. One might deduce that it can catalyze reactions involving beta-galactoside sugars. And if we think even further back to Chapter 4, RNA polymerase was the enzyme that connects ribonucleotides into a polymer.

Create Pure Oxygen! *Breakout Session 1*

Using your Blue-it Kit cell extract (after you do the lysis, centrifugation, sterilization you will have your Blue-it Kit cell extract) you will attempt to create pure oxygen. If you extract is old (more than 1 day) and has not been stored in the refrigerator, the enzymes may have lost function. In this case you will have to try again with fresh cell extract.

Have you ever had a scrape or cut in your skin that you disinfected using hydrogen peroxide? Hydrogen peroxide is very harmful to bacteria. This is why it is used as a disinfectant. K12 *E. coli* naturally create a protein enzyme called catalase (Figure 6-7). This enzyme is part of the bacteria's natural defense system. If the bacteria encounters Reactive Oxygen Species (ROS), catalase can bind to and convert those molecules into less harmful molecules.

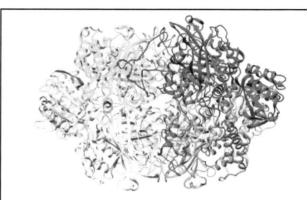

Figure 6-7. Ribbon diagram of Catalase. Catalase is a protein enzyme that can convert reactive oxygen species into less harmful molecules. It is naturally produced in *E. coli* bacteria. Source: Protein Data Bank (PDB): 1iph, illustration created with Chimera.

In this post-practice exercise, we're going take advantage of the fact that the enzyme catalase is naturally produced in *E. coli*, and that there will be thousands of catalase proteins within your Blue-it Kit cellular extract from this chapter. We're going to use catalase to convert hydrogen peroxide (harmful to *E. coli*) into pure oxygen (not harmful to *E. coli*).

The chemical reaction that catalase catalyzes is the following:

$$2\ H_2O_{2\ (l)} \text{—catalase—} > O_{2\ (g)} + 2\ H_2O_{\ (l)}$$

This means that catalase will convert two molecules of hydrogen peroxide (H_2O_2), into one molecule of oxygen gas (O_2) and two molecules of water (H_2O).

1. In a transparent container such as a beaker, drinking glass or jar, create a 1/3 solution of hydrogen peroxide with distilled water (tap water is OK, it just may not work as well). You can do this by adding 1 part hydrogen peroxide (*e.g.*, 1/3 cup) to 2 parts water (*e.g.*, 2/3 cup). You can use these measurements, or make as much or little of this mixture as you like.

2. With one of the pipet from the Blue-it Kit, transfer a small amount (~100 uL) of your cell extract into the hydrogen peroxide mixture. The small line indented on the pipet marks 100 uL.

3. After you've added some cell extract, wait a few minutes, up to a few hours, and you will start seeing magic happen. Bubbles will begin to form and rise to the surface. Bubbles will also begin to form on the edge of your container. These bubbles are pure oxygen gas that comes from the chemical reaction that the catalase enzyme is catalyzing!

[Raw material [Substrate(s)] —> Finished Product [Product(s)]

This is a very common theme among manufacturers such as large steel mills that take in raw materials such as iron ore and refine it through various processing into finished products such as steel beams that are used for building. Pharmaceutical companies use enzymatic processing to process less useful molecules into real medicine.

The catalase enzyme will complete the Four B's inside the hydrogen peroxide container. Catalase will bump around until two hydrogen peroxide molecules bump into the catalase and bind. The catalase will cause a burst, meaning it will create a chemical reaction that breaks the hydrogen peroxide molecules apart and reforms them into one molecule of oxygen and two molecules of water. Once catalase completes the chemical reaction, the water and oxygen will leave the enzyme and catalase continues to catalyze further chemical reactions. A single catalase enzyme can complete the chemical reaction thousands of times every second!

If all of the hydrogen peroxide is converted into end products, how much oxygen will be created? Let's calculate a simple approximation:

The bottle of hydrogen peroxide purchased from the pharmacy is 3% hydrogen peroxide, this means that for every 100 mL of liquid, there are 3 grams of hydrogen peroxide. Say we added 100 mL of hydrogen peroxide solution to 200 mL of water, this means we added 3 grams of hydrogen peroxide.

For this reaction, assuming the conservation of mass in a chemical reaction where the mass of the reactants equals the mass of the products, the mass of the oxygen gas (O_2) and water (H_2O) will also be 3 g. If this assumption is valid, then the total molecular weight of the reactants will also equal the molecular weight of the products. What percentage of the end products mass belongs to oxygen gas?

2 H_2O_2 —-catalase—> O_2 + 2 H_2O

In each reaction:
- 1 oxygen gas molecule has a molecular weight of 31.998 g/mol
- 2 water molecules have a combined molecular weight of 36.030 g/mol (2 molecules x 18.015 g/mol)

The total molecular weight of the products is 31.998 g/mol + 36.030 g/mol = 68.028 g/mol. This is consistent with our assumption in that the total molecular weight of the hydrogen peroxide reactants is similar (68.028 g/mol).

Of the total molecular weight of the products, what percentage is oxygen gas?

O_2 / (O_2 + 2 H_2O) x 100 = percentage oxygen molecular weight

31.998 g/mol / (31.998 g/mol + 36.030 g/mol) = 47.03 %

47.03% of 3 g is 1.41 g of oxygen gas! Let's take it one step further. How much volume will 1.41 g of O_2 occupy? The density of oxygen gas at standard temperature and pressure (STP) is 1.429 g per litre. With 100 mL of store-bought hydrogen peroxide, you can create:

1.41 g / 1.429 g per L = 0.99 L of oxygen gas.

Can you calculate how many mL of water is created by catalase during the chemical reaction? Use your cipher-skills from Table 5-2 to check your answer

Table 6-1. Be the RNA polymerase and the Ribosome to check your answer															
	5'														3'
Leading DNA (+)	aac	gaa	ccg	aac	acc	ttt	att	gtg	gaa	aac	att	aac	gaa	atg	ctg
Template DNA (-)															
	3'														5'
RNA															
Amino Acids	O														

CONGRATULATIONS!

on completing your fifth set of experiments!

Fundamentals: Diving into enzymatic processing

The basics of enzymatic chemical reactions

The basic model of a chemical reaction involving an enzyme (E), a substrate (a), and a product (b) is as follows. Recall that the substrate is the starting molecule and the product is the end material resulting from the chemical reaction (Figure 6-8):

E + a <—> Ea —> *Ea* —> Eb —> E + b

1. The enzyme (E) and substrate (a) are independently bumping around the cell (E + a). There may be hundreds, thousands or millions of each in the cell at any given time. The numbers depend on the cell metabolism and the environment the cell is in.

2. They bind because of complementary molecular shapes and chemical bonding (Ea). The "lock and key" mechanism is illustrated in Figure 6-8.

3. The chemical reaction occurs (*Ea*). The asterisks denote the energy change resulting in the chemical reaction happening.

4. Through the chemical reaction, the substrate molecule is changed into the product that is a different molecule and has a different shape and therefore chemical bonding (Eb).

5. The product has a different shape and chemical structure than the substrate. It no longer binds strongly to the enzyme, and they separate (E + b). The enzyme (E) and product (b) continue to bump around the cell to complete their functions.

A common way to describe how an enzyme and the substrate bind, is called the "lock and key" mechanism (Figure 6-8). A key (substrate) can "bump" into many different locks (enzyme binding site), but only one key has a very specific pattern that will match the lock mechanism and be able to turn. The molecule structures of the enzyme and substrate are similar to the shape of the lock mechanism and the key. However, unlike how a person has the intention to put the key into a lock, there is "no one" and no "intention" to put the substrate molecule into the enzyme. This is why the Four B's of bumping around the cell and chemical bonding are so important in causing the molecule to first encounter the enzyme, bump its way into the enzyme, and if the fit is proper and it binds, a chemical reaction can occur.

In the previous example there was a single enzyme (E) and a single substrate (a). What is far more common in chemical reactions in the cell is that there will be more than one substrate. You experienced this in Exercise 1 (isoamyl alcohol + acetyl-CoA) and Exercise 2 (X-gal + H_2O). The enzyme will take one or more atoms from one substrate and transfer it to the other substrate. This results in two different products (Figure 6-9). The same basic principles hold:

E + a + b <—> Eab —> *Eab* —> Ecd —> E + c + d

1. The enzyme (E) and the substrates (a) and (b) are independently bumping around the cell (E+a+b). There might be hundreds, thousands or millions of each at any given time in the cell.

2. They bind together because of complementary molecular shapes and matching chemical bonding (Eab). The "lock and key" mechanism illustrated in Figure 6-9.

3. The chemical reaction occurs (*Eab*). The asterisks denote the energy change resulting in the chemical reaction happening which may or may not include the enzyme transferring some atoms from one substrate to the other. This results in two new molecule products (c) and (d): Ecd

4. The products have different composition and shapes than the substrates.

5. Because the products have different shapes and chemical structure than the substrates, they no longer bind strongly to the enzyme, and they separate (E + c + d). The enzyme (E) and products (c + d) continue to bump around the cell to complete their functions.

The Four B's and enzyme function

In Chapter 4 you learned about the basic operating environment of a cell. You learned that molecules do not have intentions, nor do they have intelligence. Instead, three primary factors drive the operation and "decision making" in cells:

- The number of individual or groups of molecules.

- The rate at which molecules bump into other molecules and move around the cell.

- The strength of interaction between molecules.

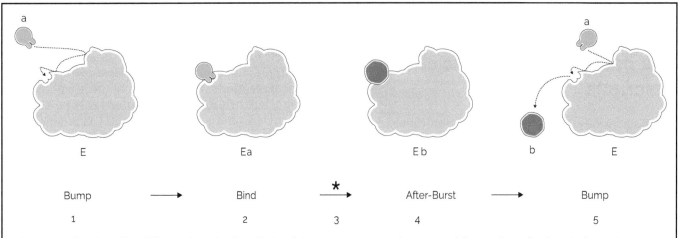

Figure 6-8. "Lock and key" illustration: the Four B's involving an enzyme, a substrate, and the product of a chemical reaction.

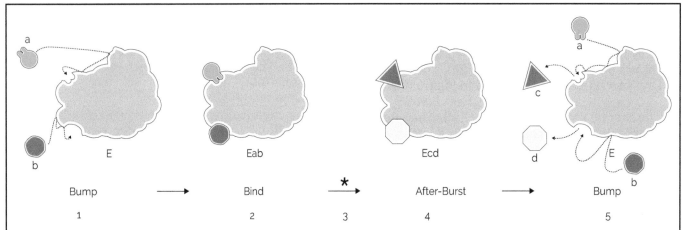

Figure 6-9. A protein enzyme can also bind to multiple substrates and process them into multiple different products. The shape of the molecule is like the shape of a key, and the shape of the binding region of the enzyme is the like the lock.

In this section, we are going to look more closely at this concept in the context of protein enzymes.

- **The number of molecules:** The number of protein enzymes and the number of substrate molecules (reactants) are essential factors in determining how frequently a chemical reaction will happen. If you have twice as many enzymes or substrate molecules in the cell, this increases the probability that they will encounter one another while bumping around.

In Exercise 2, you added extracted beta-galactosidase enzymes to your tube of X-gal substrate reaction buffer. If you had added more X-gal to the reaction, there would have been more substrate molecules bumping around and more substrate bumping into the enzymes resulting in an quicker color change. Similarly, an increased number of enzymes would increase the chance of the substrate bumping into the enzymes, resulting in a chemical reaction and a color change happening faster.

- **The rate at which molecules bump into other molecules:** The bigger a molecule is, the slower it moves around the cell. If a protein enzyme is very large, it will bump around the cell slower, and therefore encounter the substrate molecules less frequently than a small protein enzyme. Similarly, a large substrate molecule will move more slowly around a cell than a very small molecule.

So, how often does a small molecule like X-gal bump into a protein such as beta-galactosidase? While scientists haven't figured out the rate for every molecule and many factors are at play in determining the rate at which molecules move,

X-gal is considered a "normal" small molecule with a "normal" rate of movement for a small molecule. A "normal" amount can be estimated at up to 10^{20} collisions per second. In the world of biochemistry, each molecule can have up to 100,000,000,000,000,000,000 or 100 quintillion collisions with other molecules each second! If you were an atom or small molecule, this is the equivalent of you brushing elbows with every person on earth 17 billion times each second. In other words, every protein enzyme in the cell, such as beta-galactosidase, has MANY opportunities to interact with its substrate, X-gal, hundreds or thousands of times each second. This is why an enzyme can cause a chemical reaction to happen. You might recall RNA polymerase can typically add 50 ribonucleotides to an RNA transcript each second. A single catalase enzyme can turn thousands of hydrogen peroxide molecules into oxygen and water every second.

This can be simplified into knowing that molecules typically interact ("bump") a billion times a second with other molecules within the cell. This means that often two molecules will bump into one another countless times per second and it might take millions of bumps before a substrate "key" bumps into the enzyme "lock", resulting in a chemical reaction.

- **The strength of interaction between molecules:** The most important factor in determining the operation or decision making in a cell is bonding. The specific interaction between molecules is the "logic" that drives specific events in cells. In the X-gal example, the protein enzyme beta-galactosidase has the right structure and composition to be able to bind quite specifically to X-gal, and X-gal alone. It does not bind to other molecules in the cell with any notable strength or specificity. It is 'beta-galactosidase's job' to bind specifically to X-gal and catalyze the chemical reaction that cuts off a sugar ring allowing X-gal to dimerize and become colorful (Figure 6-6).

If the interaction is strong, the substrate is more likely to get "locked" in place in the protein enzyme active site, the site at which the chemical reaction happens. If the bonding interaction between the enzyme and the substrate is weak, then the substrate will not "lock" into place in the enzyme active site. The strength of an interaction between molecules is defined by its dissociation constant, discussed in the breakout below.

The same principles apply to the thousands of other protein enzymes in a cell. But what is an interaction? What does it mean when a substrate gets "locked into place"? In the next sections, we'll dig a little deeper into atoms, molecules, and the bonding that makes it all happen so that we may answer these questions.

Atoms

Before learning what causes the interactions between substrates, protein enzymes and having a more in-depth discussion about chemical reactions, we need to see what an atom is.

An atom is very small and generally considered the smallest unit of matter. A single atom is usually about 300 picometers in width, which is 0.0000000003 meters, or about a million times thinner than a human hair. An atom has two essential attributes:

- **Nucleus:** a very dense core called a nucleus, contains sub-atomic particles called protons (which are positively charged), and neutrons (which have no charge). The nucleus makes up more than 99% of the mass of each atom and has a positive charge.

- **Electron clouds:** electrons are very small negatively charged atomic particles that orbit the nucleus very fast in "cloud-like" patterns. The electron clouds are called orbitals, and they are negatively charged.

Dissociation constant *Web Search Breakout*

The equilibrium dissociation constant (K_d) is used to evaluate and rank strengths of molecular interactions. You can search for this, as well as "Enzyme Kinetics", "Michaelis-Menten Kinetics" and "Specificity Constant" online to learn more in-depth. Each of these topics discusses how the strength of bonding between a substrate and an enzyme relate to the specificity of the chemical reaction (how well a substrate "locks in place"), as well as the speed of the chemical reaction.

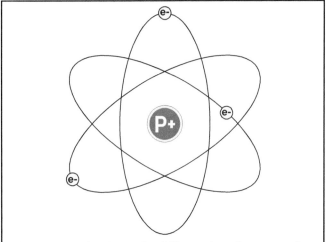

Figure 6-10. Classic out-dated illustration of an atom. Electrons (yellow) orbit the positively charged nucleus (red) like moons orbiting a planet.

The classic illustration of an atom can be found in Figure 6-10. In this illustration, negatively charged electrons were thought to orbit the positively charged nucleus of an atom in a single ring-like path like the ring around the planet Saturn. Throughout the 20th century, however, our understanding of atoms changed, and we now know that electrons do orbit the nucleus, but rather than doing so in a single ring, they make three-dimensional structures that look sort of like different balloon shapes. Scientists call these orbitals. Every electron orbiting an atom has a slightly different but predictable orbital "balloon shape" due to the size of the nucleus (number of protons and neutrons) and the number of electrons. (Figure 6-11).

Because this section is only meant to provide a basic introduction to atoms and enable you to start thinking more deeply about chemical bonding and interactions, Figure 6-11 shows you the simplest orbital paths that electrons take around a nucleus. The most simple orbital is called an "s-orbital" where up to two electrons travel around the nucleus very quickly creating a spherical balloon-like pattern (Figure 6-11 (left)). It is important to note that just like in a balloon where the rubber material makes up only a thin layer, the electron orbit is the same. The electron path does not fill the entire sphere but instead orbits it in a thin layer. In the second example in Figure 6-11 (right), called the "p-orbitals", there are three different p-orbital paths that up to three electron pairs, or six electrons, can take. Each of the paths look similar, they are simply in different orientations around the nucleus. They have what look like a "dumbbell" shape.

Further electron orbitals called "d-orbitals" and "f-orbitals" have yet further different and interesting shapes, but all orbit around the nucleus of the atom. Every one of the electrons orbiting an atom has a particular orbit and energy.

In the later sections when we will talk about chemical reactions, know that the different atoms involved in the chemical reactions give away, accept, or share electrons from these clouds. Imagine in your minds-eye these balloon-like electron orbitals forming, disappearing, and overlapping around the nuclei of atoms in molecules as chemical reactions happen.

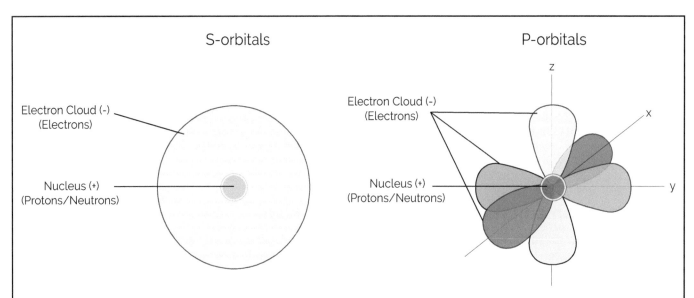

Figure 6-11. 3D Balloon-like electron clouds called orbitals. Electrons orbit the nucleus very fast and create three-dimensional patterns that look like balloons. S-orbitals are spherical, and p-orbitals look like dumbbells.

It can be difficult to understand in your mind how one or two electrons can create a "cloud". An analogy that can help is to think about the blades of a fan.

You may have noticed that when looking at a fan when it is in the "off" position you very clearly see the blades of the fan. Perhaps three blades that are very clear and distinct. In this analogy, electrons are like the blades. The electron(s) are distinct particles. When you turn on the fan, the blades spin quickly, and they appear to be a disk. It is not because the blades morph into a new form, but rather because the blades are moving fast in a circle and they are perceived to be a disk. This is similar to an electron and an electron cloud orbital. The electron moves very fast around the nucleus a spherical path (in s-orbitals) that appears to be a spherical electron cloud!

To learn more about orbitals and orbital theory you can look to Khan Academy. If you're interested in seeing how two atomic orbitals can come together to form a molecular orbital, have a look at a Youtube video where the instructor discusses what happens when two oxygen atoms come together to form a molecule of oxygen gas that you're breathing right now! https://amino.bio/molecularorbital

Bonds

Now that you have a basic idea of what an atom is, we can discuss bonding. Bonds are the "joints" between atoms. Strong bonds can hold atoms together to form stable molecules, and weaker bonds can temporarily hold atoms and molecules together. Molecules are two or more atoms. There are many kinds of bonds. Here is a high-level overview of what they are:

Covalent Bonding: These bonds are the strongest bonds, generally having a bonding energy of ~200-400 kJ/mol. Covalent bonds occur when an electron from an atom is shared with another. More specifically, when an electron in an orbital of one atom overlaps with another atom's electron orbital, the electron can become shared. This sharing of electrons creates an attraction between the atoms. For example, in a carbon-carbon bond (a common bond in living organisms - like in the sugar molecules in Figure 3-18) you'll see that the electron orbitals of one carbon can overlap with another, which results in one or more bonds (Figure 6-12).

Covalent bonds are abundant in biochemistry. These bonds bring the hydrogen (H), carbon (C), nitrogen (N), oxygen (O), phosphorous (P), and sulfur (S) atoms together to form all of the biomolecules in cells. Proteins, DNA, RNA, lipids, and sugars molecules are made of CHOPNS and are all held together by covalent bonds, which you can learn more about in Table 6-1.

What does "a bond energy of ~200-400 kJ/mol" mean? The basic unit of energy is a joule (J), named after the famous scientist James Prescott Joule who defined

the joule. The joule is used to measure the amount of energy in the world around us, including in chemical bonds. For example, as a general rule, the thermal heat energy available at room temperature is 3,000 joules per mole or 3 kilojoules per mole (~3 kJ/mol), which is much lower energy than the typical covalent bond energy of 200-400 kJ/mol. In other words, all of the matter around you that make up chairs, desk, table, and even the gas that you're breathing have a heat energy of about 3 kJ/mol. What is a mole? Going Deeper 6-5 will help you dive into this topic.

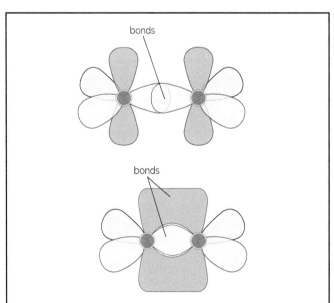

Figure 6-12. Orbitals in two carbon atoms can overlap to form bonds. On the top there are two carbon atoms bound because of the overlapping yellow dumbbell shaped orbitals - this results in a single bond between the carbon atoms. Under the right conditions, the green "vertical" orbital can also overlap to create a second bond (bottom).

The next time you eat a chocolate bar, have a look at the nutritional information to see how much sugar there is in it - you might be surprised! Let's assume for this discussion that it has 10 grams of sugar in it. Of what use to scientists is "10 grams of sugar"? It turns out that simply knowing the mass of something does not tell you very much. What is much more helpful is knowing how many sugar molecules there are. Look back at Figures 6-8 and 6-9 and consider the enzyme reaction. Is it more useful to know that there are 10 substrate molecules per one protein enzyme or 0.00000001 g of substrate and 0.0000001 g of enzyme? In chemistry, the number of molecules and ratios of molecules provides you with more information than just the mass. This is in part because molecules are different "sizes" and so one gram of sugar does not have the same number of molecules as one gram of salt. Inventing the mole was a way to calculate the number of molecules of any substance from the mass so they could be compared and more accurate calculations could be completed.

In the 1800s, scientists wanted a way to find the equivalent number of atoms/molecules in different substances using their mass. Using chemistry, scientists determined a very close approximation of the number of carbon atoms in 12 grams of pure carbon to be 6.022×10^{23} atoms. That is 602,200,000,000,000,000,000,000 carbon atoms in 12 grams of pure carbon. They then decided that a mole would be 6.022×10^{23} atoms or molecules of any pure substance. With this knowledge, chemists then determined how many grams of each chemical element is needed to get 6.022×10^{23} atoms (one mole). This became known as the atomic weights. If you look at the periodic table at the end of this book, you'll see the atomic weights of the elements. Hydrogen, for example, has an atomic weight of 1.008 g/mol. This means that in 1.008 grams of hydrogen, there are 6.022×10^{23} atoms, 1 mole, of hydrogen atoms.

You can also combine atomic weights to find the molecular weights. Water (H_2O) has one oxygen (15.999 g/mol) and two hydrogens (1.008 g/mol x 2 = 2.016 g/mol). Add the **atomic weights** up, and you get a **molecular weight** of 18.015 g/mol. This means that 18.015 grams of water, 1 mole, has 6.022×10^{23} water molecules! Next time when you drink a 500 mL bottle of water, you can boast that you just drank:

500 mL water is *500 g water, divide by, 18.015 g/mol = 27.75 mol*

27.75 mol water x 6.022×10^{23} molecules per mol = **1.67×10^{25} water molecules!**

(That is 16,700,000,000,000,000,000,000,000 molecules of water in a 500 mL bottle!)

Table 6-1. Enthalpies and lengths of different bonds					
Bond	**Length (pm)**	**Energy (kJ/mol)**	**Bond**	**Length (pm)**	**Energy (kJ/mol)**
H-H	74	436	C–N	142.1	305
H-C	106.8	413	C=N (double)	130	615
H-N	101.5	391	C≡N (triple)	116.1	891
H-O	97.5	467	C–O	140.1	358
C-C	150.6	347	C=O (double)	119.7	745
C=C (double)	133.5	614	C≡O (triple)	113.7	1072
C≡C (triple)	120.8	839	O–O	148	146
C-S	182	272	O=O (double)	120.8	498

Source: http://philschatz.com/chemistry-book/contents/m51056.html

Now that you have a better idea of what a mole is (Going Deeper 6-5) let's get back to the kJ/mol of heat energy. This means that a mole of substance around you, such as one mole of oxygen gas in the room around you, or one mole of molecules in the chair you're sitting on, has heat energy of around 3 kilojoules. However, in one mole of atoms bound together via a covalent bond, the bond energy holding the atoms together is 200-400 kilojoules (kJ/mol).

To break a bond, you need to add at least the energy of that bond. For example, ~200-400 kJ/mol of energy needs to be added to a molecule in order to overcome and break the covalent bonds. Because normal room temperature energy is ~3 kJ/mol, all bonds with higher than 3 kJ/mol energy are stable at room temperature because there is not enough thermal energy (heat) to break them. This is, for example why the graphite in your pencil stays stable - the carbon-carbon bonds that make up the graphite have a bond energy of around 347 kJ/mol (single bond) and 614 kJ/mol (double bond). In the coming sections, you'll soon learn about some other types of bonds that are affected by normal room temperature energy and are a reason why butter may melt if left out of the fridge on a warm day and why the heatshock step during a transformation makes the membranes of cells more fluid.

Because most chemical reactions that happen in cells involve the creation or breaking of covalent bonds, which have much higher bond energy than the 3 kJ/mol energy available at room temperature, a catalyst is needed to replace the need for such energy. As you saw in the Going Deeper 3-7 on chemical reactions of Chapter 3, a catalyst is a substance that lowers the activation energy needed to cause chemical reactions to happen. In the case of biology, the catalysts that lower the activation energy to break covalent bonds, are protein enzymes, such as the beta-galactosidase and ATF1 protein enzymes you genetically engineered the *E. coli* cells to produce. Rather than needing the "height off the ground" 327 kJ/mol energy to break a carbon-carbon bond, the enzyme helps create a "tunnel" to make the reaction happen without the excess energy.

Protein enzymes bind to substrate molecules, twist them, bend them, and even bring substrate molecules into close proximity to "force" chemical reactions to happen. In some instances, the amino acids that make up the protein enzyme have extra electrons that will kick-start the chemical reaction by forming bonds with the substrate. When a protein enzyme does this, it changes the rules of the game and it lowers the energy required to break and form bonds and make the chemical reaction happen.

This is the magic of living systems. Without protein enzymes, very few chemical reactions would happen because covalent bonds are quite stable. Life would not exist without enzymes. It is the thousands of protein enzymes in the cell that catalyze specific chemical reactions to happen and sustain life.

Have a look back at the various chemical diagrams of the different macromolecules - nucleic acids, lipids, sugars, and proteins. You'll see that all of these important molecules, made up of CHOPNS, are joined together by covalent bonds. Thousands of molecules are integral to life and are made up of an assortment of covalently bonded CHOPNS atoms. All of the atoms are joined together by the sharing of electrons in their balloon-like orbits (orbitals).

Electron orbitals and valence shells *Going Deeper* **6-6**

The outer most electron orbital in an atom is called a 'valence shell'. Some atoms have their outer most valence shells almost full or almost empty. To become more stable, atoms have natural propensity to be full or empty and to do this, atoms can gain or lose electrons. Alkali metals such as lithium (Li), sodium (Na), and potassium (K) all have one valence electron in their outer s-orbital. To become more stable, they prefer to lose this to another atom. When they lose an electron, they get a positive charge. This is why you typically see Li^+, Na^+, and K^+, these are the most stable forms of those atoms.

Alkali metals are well known to lose their single valence electron to halides such as fluorine (F), chlorine (Cl), bromine (Br), and iodine (I). This is because the halides have an almost full valence p-orbital shell and they would like to fill it up with one more electron to become more stable.

As an example of this, when sodium (Na) and chlorine (Cl) are combined, the sodium will spontaneously transfer an electron to chlorine to become sodium (Na^+) and chloride (Cl^-). Now that the atoms have become charged ions, they participate in ionic bonding. Search "valence shell" online to learn more.

Ionic Bonding: Electromagnetism, a well-understood field, describes how positive charges repel, negative charges repel, but a positive and a negative charge attract. In cells, there are many cases where positively charged atoms called cations ('cat-ions'), and negatively charged atoms called anions ('an-ions') form an interaction due to the rules of electromagnetism (Figure 6-13). This is called an ionic bond. Ionic bonds typically have an energy of ~30 kJ/mol - 10 times less strong than covalent bonds. Ionic interactions are long-range interactions, as compared to covalent interactions that require two atoms to be so close that their electron orbitals overlap and electrons are shared. In other words, ionic bonds do not bring atoms together to form molecules, rather they can be the temporary bonds between atoms and molecules.

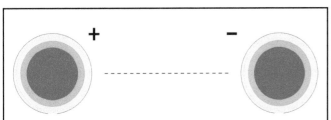

Figure 6-13. Positively charged and negatively charged ions are attracted to one another by electromagnetic force. The dashed line indicates an attraction interaction.

You learned about calcium in Chapter 4 when you completed your transformation experiment. The positively charged cationic metal calcium (Ca^{2+}), is able to interact through ionic bonding with the DNA plasmids and the outer cell membrane. This is because the DNA has a strongly negatively charged backbone partially comprised of phosphate (PO_4^{1-}). You also learned about the 20 standard amino acids, and how each of them has a unique side chain. If you look back at Figure 3-30, you'll see that some of the amino acids have a positive charge, negative charge, and some are uncharged. Ionic bonding is key in driving some of the interactions that cause the proteins to fold up into larger 3D structures (Figure 3-28).

Ionic interactions are also important in causing molecules to begin interacting with one another from a distance. Recall that during the Four B's, the first being bump, molecules will bump around until they bind. Ionic interactions can pull two bumping molecules together during the bumping phase. For example, ionic bonding can help the substrate "key" find the enzyme "lock" from a distance. It is important to note that unlike how electrons are actually shared between atoms in covalent bonding, ionic bonds do not result from electron sharing. Ionic bonding is caused by the electromagnetic forces of positive or negative charges of atoms and molecules attracting or repulsing.

Hydrogen Bonding: This is the third strongest class of bonds generally with 5 - 30 kJ/mol bond energy. Hydrogen bonds are similar in nature to the ionic bonds you just learned about. Hydrogen bonds are weak ionic bonds created when a hydrogen atom (H) that is covalently bound to an electronegative atom such as a nitrogen (N), oxygen (O), or fluorine (F), is attracted to another negatively charged atom.

Electronegative atoms are those that have a strong desire for electrons, and they have the ability to tug on the electron s-orbital in a hydrogen atom that is covalently attached to them. This results in the electron orbital cloud around the hydrogen atom being slightly positively charged because the positively charged nucleus is no longer fully shielded by its s-orbital (Figure 6-14).

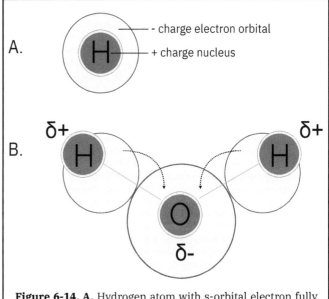

Figure 6-14. A. Hydrogen atom with s-orbital electron fully orbiting the nucleus. **B.** Water molecule demonstrating electronegative oxygen tugging on hydrogen atoms' s-orbitals leading to a hydrogen bond opportunity because of the (+) charge from the nucleus.

A good example is in a water molecule (Figure 6-14), where there is one electronegative oxygen atom covalently attached to two hydrogen atoms. Normally the negatively charged s-orbital of the hydrogen fully surrounds positively charged nucleus resulting in hydrogen having no charge (Figure 6-14 A). However, in the case of when hydrogen atoms are covalently bonded to an electronegative oxygen (water), the oxygen atom pulls the hydrogen's electron toward it (Figure 6-14 B). That's right, oxygen is greedy for electrons! This means that the positively charged nucleus of the hydrogen is no longer fully surrounded by its negatively charged s-orbital and so, exhibits a partial positive charge (ϱ+). Because the oxygen atom now has a "little extra electron" from the hydrogens, the oxygen has a slightly negative charge (ϱ-).

The partial positive charge of the hydrogen atoms can now interact with the negative charges in other molecules, just like you would observe with an ionic bond. Because these are only "partial charges", meaning not "full plus" or "full minus" charge, they are not as strong as ionic bonds.

It is very common for the hydrogen of one water molecule to want to "hydrogen-bond" to the oxygen atom of another water molecule. Hydrogen bonding is highly prevalent in proteins and is the primary force that causes proteins to fold into their three-dimensional shape! Remember back to Chapter 1 when you learned two DNA strands can come together to form the double helix - it is hydrogen bonding that is important in the complementary pairing of nucleotides (Figure 6-15). It is hydrogen bonding between hydrogen-oxygen or hydrogen-nitrogen that cause DNA to zip up into a double helix. Further, in Chapter 4 when you learned about transcription, it is hydrogen bonds between the RNA transcript and the DNA (-) strand that hold RNA polymerase locked-in to the DNA. Lastly, in Chapter 5 when you learned about translation, it is hydrogen bonding that causes the tRNA's anticodon to complement with the RNA transcript codon inside of the ribosome, enabling translation to happen. Hydrogen bonding is an essential kind of bond that allows The Four B's of Cell Operation to occur.

to 4 kJ/mol. As you learned earlier, the basic thermal energy at room temperature is around 3 kJ/mol, and so these interactions can be affected by the temperature of the environment.

Have you ever heated up glue in a glue gun, or steamed a stamp to pull it off of the envelope? When doing this, you are adding enough energy in the form of heat to overcome the Van der Waals interactions and London Dispersion Forces that are holding the glue molecules together.

Just as hydrogen bonds were a weaker form of an ionic bond because they involve partial charges, Van der Waals interactions and London Dispersion Forces, follow a similar principle and are even weaker. In a hydrogen bond, the electron orbital around the hydrogen atom is distorted by a nearby atom such as another hydrogen, resulting in a positive charge on one side of the hydrogen that can interact with other (-) charged atoms (Figure 6-14). In the case of Van der Waals interactions and London Dispersion Forces, the electron orbitals in atoms aren't perfectly spherical and naturally fluctuate. Also, sometimes it just happens that there are more electrons on one side of the atom than the other. When this happens, one side of the electron orbital cloud is slightly more negative in charge than the other side. Conversely, the other side is slightly more positively charged. As you probably already guessed, this means that the slightly negative side of the atom can interact with a slightly positive side of another atom (Figure 6-16).

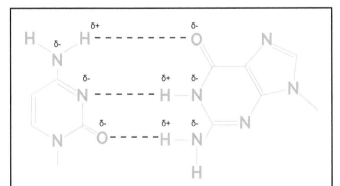

Figure 6-15. Hydrogen bonding between C-G nucleotides. Electronegative nitrogens cause hydrogen atoms to become partially positive charged, enabling hydrogen bonding with partially negatively charged nearby oxygens or nitrogens.

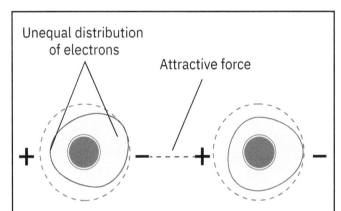

Figure 6-16. At very close proximities the fluctuation the electron density of an atom can result in one side being more or less charged than the other. Similar to an ionic bond or hydrogen bond, the slight negative charge of one atom can then interact with the slight positive charge of another and cause an interaction.

Van der Waals Interactions / London Dispersion Force: The weakest of the intermolecular forces in cells, these usually have bond strengths between 0.4

This type of bond only happens when atoms come into very close contact with one another, less than 0.6 picometers. However, if the atoms become too close (less than 0.4 picometers), then repulsion between the electron orbits occur and the atoms are repelled by the negative charge of the electron of their orbitals.

When heat is added, the thermal energy causes the electrons in the orbitals to be more lively. This makes the electrons orbit more uniformly around the nuclei of the atoms which breaks the +/- attractions. Consider when you leave butter in the refrigerator. The butter takes on a solid form because Van der Waals interactions can occur at low temperature. If you leave butter out in a hot room, the butter softens and even melts. The room's thermal energy causes the electron orbitals in the butter molecule to become more lively, which breaks the Van der Waals interactions allowing the butter to flow freely. So next time you can't butter bread smoothly because the butter is too solid, you'll know who to blame!

Van der Waals interactions and London Dispersion Forces are important in non-charged hydrophobic interactions like those that happen in the phospholipid membrane in bacteria. During the transformation in Chapters 4 and beyond, you'll recall that you first kept your cells on cold and then you increased the temperature to 42°C to assist the DNA in getting inside the cells, during what is called a heatshock. When you increase the temperature, you are increasing the thermal energy of the environment, and this begins to overcome the Van der Waals interactions and London Dispersion Forces that make the lipids in the membrane stick together tightly and be rigid. When heated, the uncharged tail groups of the lipids slip and slide more freely, and the membrane becomes more fluid, allowing DNA plasmids to enter more frequently.

If there is one key takeaway from this section, it is that electrons of atoms drive the world around us. In covalent bonds, it is the overlap of electron orbitals that drives really strong bonds. In ionic bonds, it is excess electron(s) or lack of electron(s) that makes atoms positively or negatively charged and drives the electromagnetic interaction. In hydrogen bonding, it is the tugging of the hydrogens electron orbital that causes it to be slightly positive and enables it to bind to something with a negative charge. In Van der Waals interactions it is the natural fluctuation of electron orbitals in atoms that can cause very weak positive/negative charge interactions at very close range.

As you will now see, it is also the movement of electrons that are at the heart of chemical reactions.

Protein enzyme catalysis in cells

Now that you have a much broader understanding of what an atom is, what bonds are, and the basic mechanism of enzymatic chemical reactions, we can have a more in-depth look at one of the chemical reactions that occurred when you transformed your K12 *E. coli* cells with the DNA plasmid. Let's look at chloramphenicol acetyltransferase (CAT), the enzyme that causes chloramphenicol resistance in the genetic engineering experiments done throughout this book.

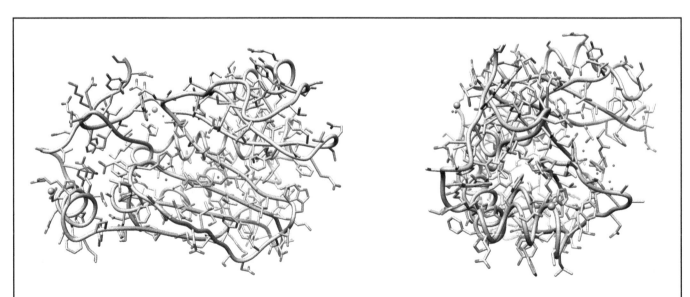

Figure 6-17. Chloramphenicol acetyltransferase bound to chloramphenicol (orange). Source: Protein Data Bank (PDB): 3cla; edited using open source software, Chimera.

When the chloramphenicol molecule is processed by CAT, it is no longer active and no longer able to harm the cells.

In Figure 6-17 the CAT protein (beige) is interacting with chloramphenicol (orange). Chloramphenicol is in the "binding pocket" or active site of CAT and held in place with hydrogen bonds formed between the amino acid side chains in the protein, water molecules, and chloramphenicol (Figure 6-18). Each of the thin blue lines indicates a hydrogen bond. If you've forgotten what amino acids look like, check back to Figure 3-30.

In Figure 6-18, the amino acid side chains of CAT (pink) are also able to bind to a second molecule called acetyl-CoA (not shown). Remember acetyl-CoA? This molecule is also an essential part of the Atf1 reaction for making overripe banana smell!

The chemical reaction that CAT catalyzes involves the amino acid histidine (Figure 6-18 (pink)), which is an important part of initiating the chemical reaction (Figure 6-19):

(a) A nitrogen atom in a histidine amino acid (pink) uses some "spare electrons" to "steal" a hydrogen from the chloramphenicol. This is shown by an arrow from two red dots (electrons) "reaching out and stealing" the hydrogen from the chloramphenicol molecule (green);

(b) The electrons that previously formed the bond with the now "stolen" hydrogen, form a new bond with the carbon atom of the acetyl-CoA (blue), which is also bound in the enzyme binding pocket. This is shown by the arrow continuing to the carbon in acetyl-CoA;

(c) Step (b) causes the bond from the double carbon-oxygen in acetyl-CoA to be broken, and the electrons move to the oxygen;

(d) The electrons only stay temporarily on the oxygen and quickly move back to reform the carbon-oxygen double bond;

(e) The movement of electrons in (c) and (d) cause the carbon-sulfur bond electrons to move to the sulfur atom of the CoA molecule, 'destroying' the carbon-sulfur bond;

(f) The acetate group is free and separate from the "CoA". The carbon that was previously bound to the sulfur of the CoA forms a stable covalent bond with the available oxygen on the chloramphenicol molecule.

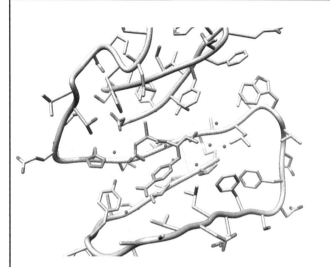

Figure 6-18. Zoom in of the chloramphenicol binding pocket of CAT and chloramphenicol. The blue lines are hydrogen bonds. The hydrogen bonds holding chloramphenicol in the binding pocket are between the amino acid side chains, water molecules (red spheres), and the chloramphenicol molecule. The histidine amino acid in the chemical reaction is in pink.

Several factors contribute to the CAT/chloramphenicol chemical reaction happening. In one instance, the substrates chloramphenicol and acetyl-CoA are put into very close proximity with one another by bumping around the cell and then binding simultaneously to CAT. The hydrogen bonding formed between the substrates and CAT hold the substrates in place, but they also have an influence on the bonds (electron orbital clouds) in the substrates which changes electronic structure of the substrates. The histidine amino acid of CAT is ideally positioned to stir up trouble and start the chemical reaction by using its extra electrons to steal the hydrogen from chloramphenicol, which starts a chain reaction of 'electron movement' events. These are themes that occur in protein enzymes.

In many chemistry and biochemistry textbooks, you will learn about how enzymes "lower the activation energy" required for a chemical reaction to occur. This is another way of saying that to break a covalent bond which has 300 kJ/mol energy you typically need 300 kJ/mol of energy. However, enzymes are able to lower the energy required to break the bond. Enzymes "change the rules" by employing some of the mechanisms you just learned about. If you haven't already, go back to the Chemicals Reaction **Going Deeper 3-7** which makes the analogy between a roller coaster and a chemical reaction. In this instance, the ability of CAT to initiate a chain of electron jumping events is the "tunnel" in the Going Deeper analogy (Figure 3-29). These electron jumps cause the creation and breaking of bonds without the addition of 300 kJ/mol energy.

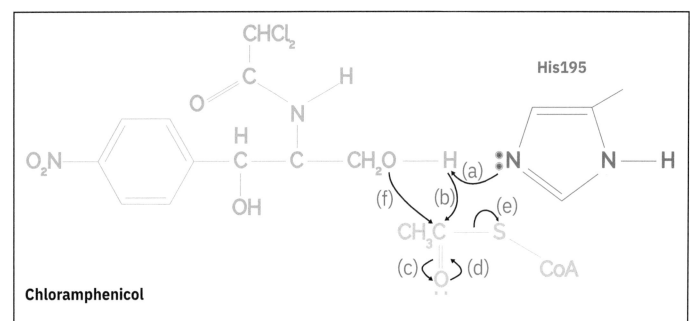

Chloramphenicol

Figure 6-19. The CAT/chloramphenicol chemical reaction includes several steps that involve the movement of electrons. The arrows in this diagram indicate where electrons, and therefore bonds, move. Chloramphenicol (green); histidine amino acid of CAT (pink); acetyl-CoA (blue). When the acetyl (blue) is connected to chloremphenicol (green), it no longer has antibiotic properties.

Atf1 *Video Breakout*

Now that you completed the Enzymatic Processing hands-on exercise and have learned about enzyme catalysis, have another look at the video covering the chemical reaction to create banana smell. You'll have a deeper appreciation for the breaking and making of covalent bonds by Atf1. Visit: https://amino.bio/atf1

X-ray Crystallography *Going Deeper* **6-7**

You might be wondering where the protein structure illustrations in Figure 6-17 and Figure 6-18 come from and why acetyl-CoA is not in Figure 6-17. The illustrations in these figures are based on real data! But how can scientists possibly take pictures of proteins at the atomic scale? How can you zoom in that far?

Scientists use a technique where they first create proteins through genetic engineering, much in the way that you did with the various proteins you engineered your K12 *E. coli* to produce. They then purify the proteins, and treat them to different chemical conditions and dry the samples out so that the proteins form crystals. Crystals are materials that are made up of repeating patterns. Scientists then shoot X-rays through the crystals and look at how the X-rays are scattered and diffracted by the proteins in the crystals. The diffraction patterns tell scientists where the different atoms of the proteins are, and using software like Chimera, you can visualize the atoms and the molecules (https://www.cgl.ucsf.edu/chimera/). This technique is called X-ray Crystallography, and it is what Rosalind Franklin used to get one of the best X-ray images of DNA, which led to the discovery of its structure in 1953.

There is no acetyl-CoA in the illustrations because no scientist has been able to successfully create crystals that have CAT, chloramphenicol, and acetyl-CoA. While you might find it easy to create water crystals by simply putting water in the freezer, it can take years for scientists to formulate the right experimental conditions to cause the proteins and the substrates to form crystals!

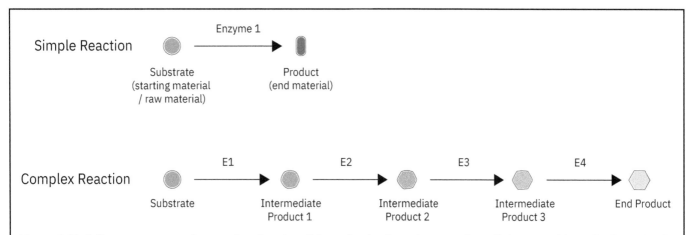

Figure 6-20. Cells process many substrates into "products" through microfacturing, more formally known as biosynthesis. In simple reactions, a substrate is turned into a product with a single enzyme (E). In complex pathways, several enzymes (E) are involved, and the products of one enzyme reaction become the substrates of another enzyme reaction.

Enzyme processing in cells: Amino acids

You've been learning about enzyme processing in the context of enzymes that you've engineered the cells to produce. However, as has been mentioned, cells already rely heavily on protein enzymes for normal operations. Let's take a quick look at proteins and the amino acids that they are made up of in the context of enzymatic processing. Where do amino acids come from? From enzymatic processing!

Using the *Three Steps to Microfacturing*, *E. coli* cells complete a series of enzymatic reactions to modify sugars and other small chemical molecules into amino acids. Another name for this is biosynthesis.

In the exercises in this chapter you completed simple chemical reactions (Figure 6-20, top) where one or two substrates were converted into products by a single enzyme. However, cells often do complex chemical reactions that require more than one enzyme to cause a chain of chemical reactions where the product(s) of one enzyme reaction becomes the substrate(s) of another (Figure 6-20, bottom). *E. coli* cells use complex enzymatic reactions to create amino acids.

In Figure 6-21 you will see a real amino acid biosynthesis map which shows many different chemical reactions. Within *E. coli* cells, all chemical reactions in the map happen simultaneously and result in the creation of amino acids. At first glance, it looks very complicated but, when you understand how it works, it's quite simple. As simple as using Google Maps to get directions from Point A to Point B:

- The open circles are a molecule, like a sugar, a fat, or amino acid (not proteins).

- The arrows represent a chemical reaction and its direction. The molecule at the base of the arrow is converted into the molecule at the tip of the arrow.

- In general, the arrow also represents a protein enzyme that causes that chemical reaction to happen.

Find a molecule in Figure 6-21 called Fructose-6P (Fructose-6-Phosphate) on the top left. Where have you heard of fructose before? You've probably heard of high fructose corn syrup, the sugar that comes from corn and is used to sweeten food in North America. Corn syrup is one form of fructose that can be found in candy, soda pop, Slurpees, and a lot more. Bacteria also enjoy sweet things. In fact, they use sugar from their environment, some of it being fructose, to make amino acids. For example, the *E. coli* in your large intestine will directly use some of the fructose from your candy and turn it into amino acids your body needs - perhaps you can use that as an excuse to eat more candy! In the case of the Figure 6-21, Fructose-6-Phosphate is a fructose sugar molecule that was taken into a cell and slightly modified by enzymes called fructokinase or hexokinase to have a phosphate added to carbon C6 of the sugar molecule. Fructose-6P can be converted to many other molecules by many different enzymes, including being converted into amino acids.

Using Figure 3-30, you can correlate the structure of the amino acids within the biosynthesis pathway in Figure 6-21. You can see that it takes many chemical reactions to convert fructose-6P to an amino acid.

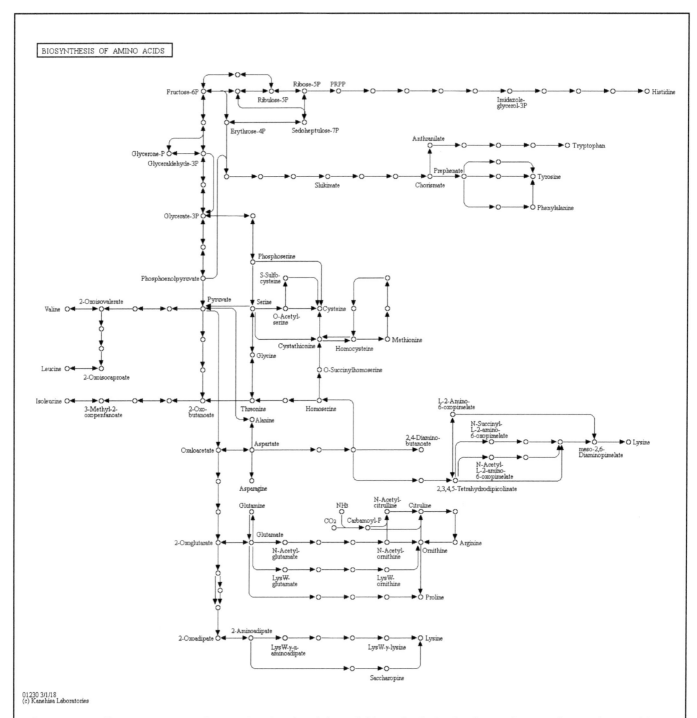

Figure 6-21. Cells process many substrates into 'products' through biosynthesis. In simple reactions, a substrate is turned into a product. In complex pathways, the products of one enzyme reaction become the substrate of another enzyme reaction. Image Copyright Kyoto Encyclopedia of Genes and Genomes (https://www.genome.jp/kegg/)

Notice that some of the open circles in Figure 6-21 don't have names to them yet. However, once you complete the Exploring Metabolic Pathways Online Websearch Breakout, you will be able to explore the biosynthesis pathways online.

In Table 6-2 is the step by step metabolic process for

converting the sugar D-fructose to the amino acid L-histidine. Each line of the table can be read as:

"In step ___ of the histidine amino acid biosynthesis pathway, the substrate molecule _____ is converted to the product molecule _____ by the protein enzyme _____".

Step #	Starting molecule (substrate)	End molecule (product)	Protein enzyme catalyzing the reaction
		Table 6-2. Non-coding regions that function in DNA and RNA	
1	D-fructose	D-fructose-6P	hexokinase
2	D-fructose-6P	D-arabino-3-Hexulose 6P	6-phospho-3-hexuloisomerase
3	D-arabino-3-Hexulose 6P	D-Ribulose 5P	3-hexulose-6P synthase
4	D-Ribulose 5P	D-Ribose 5P	ribose-5P-isomerase A
5	D-Ribose 5P	5-Phosphoribosyl diphosphate	ribose-phosphate pyrophosphokinase
6	5-Phosphoribosyl diphosphate	1-(5-Phosphoribosyl)-ATP	ATP phosphoribosyltransferase
7	1-(5-Phosphoribosyl)-ATP	1-(5-Phosphoribosyl)-AMP	phosphoribosyl-ATP pyrophosphohydrolase
8	1-(5-Phosphoribosyl)-AMP	Phosphoribosyl-formimino-AICAR-phosphate	phosphoribosyl-AMP cyclohydrolase
9	Phosphoribosyl-formimino-AICAR-phosphate	Phosphoribulosyl-formimino-AICAR-phosphate	phosphoribosylformimino-5-aminoimidazole carboxamide ribotide isomerase
10	Phosphoribulosyl-formimino-AICAR-phosphate	D-erythro-Imidazole-glycerol 3P	cyclase
11	D-erythro-Imidazole-glycerol 3P	Imidazole-acetol phosphate	imidazoleglycerol-phosphate dehydratase
12	Imidazole-acetol phosphate	L-Histidinol phosphate	histidinol-phosphate aminotransferase
13	L-Histidinol phosphate	L-Histidinol	imidazoleglycerol-phosphate dehydratase
14	L-Histidinol	**L-Histidine**	histidinol dehydrogenase

Exploring metabolic pathways online *Web Search Breakout*

Exploring metabolic pathways online: Biosynthesis maps are an essential part of engineering cells to produce molecules for us. The Kyoto Encyclopedia of Genes and Genomes (KEGG) is a great resource to find molecules that are made biologically, to understand the chemical reactions in cells that result in the production of the molecules, and to ultimately discover what DNA sequence(s) in the cells blueprints encode for the proteins that cause the chemical reactions. Try searching "Kegg fructose amino acid synthesis" on a search engine to find the biosynthesis map 01230 in Figure 6-21.

Summary and What's Next?

Congratulations, in this chapter you dug deep into the world of chemistry and learned protein enzymes catalyze chemical reactions by reducing the amount of energy required to make or break bonds resulting in substrates being processed into products. This is what we call Enzymatic Processing, which is the third step and final step of the *Three Steps to Microfacturing*.

You learned about how an atom is made up of a dense positively charged nucleus and electron clouds called orbitals around the nucleus. The electrons in the outer valence orbitals can overlap with those of other atoms to create covalent bonds. You also learned about how electromagnetic bonding is present at different levels in ionic bonding, hydrogen bonding, and even Van der Waals interactions and London Dispersion Forces. You saw that while covalent bonding is critical in holding atoms together to form molecules, hydrogen bonding is critical to hold molecules together loosely. Hydrogen bonding is key for holding DNA strands together, helping proteins interact with DNA, for substrates binding to enzymes, and a lot more.

Lastly, we looked into an actual reaction mechanism in which an enzyme catalyzes changes in molecules. CAT binds acetyl-CoA and chloramphenicol in a reaction in which CAT initiates a chain of electron jumping events that causes an acetate group to be transferred from the acetyl-CoA to the chloramphenicol rendering the antibiotic inert and unable to harm the cells.

These *Fundamentals* were reinforced by the hands-on exercises you completed where you completed enzymatic reactions involving one or more substrates. In those reactions, you processed small molecules to change odor, change color and even create oxygen.

You've only touched the surface of the world of chemistry and bonding, and as you continue your journey into genetic engineering beyond this book, you will be learning more about this subject. Mastering the concepts of chemistry and bonding are essential to mastering your abilities to manipulate cells through genetic engineering.

In the next chapter, you are going to look deeper into genetic regulation. Up until now the DNA plasmids you have used were completely autonomous, meaning once they were in the cells, they did everything automatically. In the next chapter, you're going to learn how you can turn genes on and off using different mediums such as chemicals, temperature, and light.

Review Questions

Hands-on Exercise

1. Why does one waft?

2. What is the difference between the two negative control plates in the Smell-it Experiment?

3. What are the experiment steps to get you to microfacture banana smell molecules?

4. What made it feasible to complete the Blue-it Kit using cell extract rather than in the cells (like the Smell-it Kit)

5. What is DTT? Why is it used in the lysis and reaction buffers?

Fundamentals

1. Explain the Four B's in relation to a one or two substrate-enzyme reaction.

2. How is the classic illustration of an atom incorrect? (Figure 6-10)

3. What is a mole?

4. Describe how a single electron can create an orbital.

5. What shapes do s-orbitals and p-orbitals have?

6. What are the four major types of bonding?

7. What type of bonding is used when two strands of DNA zip together to form the double helix? Explain.

8. How does chloramphenicol acetyltransferase start the chemical reaction that is important for selection during genetic engineering?

9. What is the difference between a simple enzyme reaction and a complex enzyme reaction?

Chapter 7

Manually turning on genes *in situ*

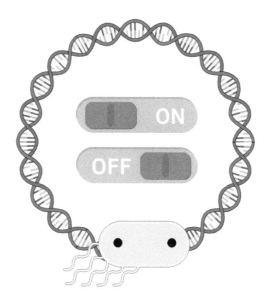

Congratulations! You are well on track to becoming a Genetic Engineering Hero! Your ability to complete fundamental techniques like making LB agar plates, transforming *E. coli*, incubating cells and even extracting enzymes and proteins are reaching post-graduate university levels. In fact, many graduate students have never made LB agar plates! In this chapter, we're going to build on the information and skills you've learned and practiced throughout the book to begin controlling *E. coli* cells after they have been genetically engineered.

Up until now, you have been engineering cells with plasmids that automatically operate. The cells automatically copied the plasmids, and the genes automatically turned on. These features were built-in to ensure the hands-on exercises and *Fundamentals* were within grasp.

This approach enabled you to complete the *Fundamentals* sections in Chapters 4 and 5, so you could understand how transcription and translation start, run, and stop. You may want to jump back to review those topics as they are essential to understand once you start manually operating genes and become a genetic engineer.

In this Chapter, you are now going to complete three different exercises teaching you how genes can be turned on manually using various mediums such as chemicals, temperature, and light. The technical term for turning on genes manually is 'inducing expression'. Inducing expression is an instrumental skill to add to your genetic engineering toolkit.

In the *Fundamentals* section, we're going to look deeper into how gene regulation works. While you will not be creating your own DNA plasmids, this chapter will provide you with some deep insights into what you should be thinking about when you do start designing and building your own plasmids and genes. We will relate what you learned in Chapters 4-6 to how the "automatic gene regulation" works and then look at how manual gene regulation works.

Getting Started
Equipment and Materials

The **Amino Labs Light-it Kit™, Heat-it Kit™,** and **Induce-it Kit™** are part of the ***Zero to Genetic Engineering Hero Kit Pack Ch. 5 - 7.*** These kits can also be ordered separately at https://amino.bio/products

Shopping List *Choose one of, a combination of, or all exercises:*

Exercise 1:
Wetware kit: Amino Labs Induce-it Kit™ (https://amino.bio)

Exercise 2:
Wetware kit: Amino Labs Heat-it Kit™ (https://amino.bio)

Exercise 3:
Wetware kit: Amino Labs Light-it Kit™ (https://amino.bio)
The Light-it-enabled DNA Playground Large or the standalone Light-it LED add-on

Exercise 1, 2 & 3:
Minilab (DNA Playground)
Microwave

Instructions Overview

Day 1-4
1. Complete the Engineer-it Kit exercise(s) as required to obtain engineered cells. (Note: The Light-it Kit does not require engineering of cells).

Day 4-5
2. Culture your cells and manually turn on gene expression:

   A. Chemical: Dissolve the inducer chemical using the dissolving buffer provided in the kit. Mix it well and add the dissolved inducer to a region of the plate. The inducer will dissolve into the LB agar. The area where you added the liquid drop will be the most induced and neighboring regions will also be induced, just not as much. You will see a "gradient" of induction. You can put drops of inducer in multiple locations.

   B. Temperature: Place your engineered bacteria in the DNA Playground incubator initially at 30 °C Then shift the temperature to 42 °C causing the cells to maximally express your trait.

   C. Light: The color of light needed to induce expression is quite specific. Turn on the desired color using the sliders on the DNA Playground Large or by using the standalone Light-it LED add-on.

Chapter Timeline Overview

The timeline to complete exercise 1 & 2 is:

Day 1: ~60 minutes with 12-24 hours incubation,
Day 2: ~60 minutes with 12-24 hours recovery
Day 3: ~30 minutes with 24-48 hours incubation
Day 4: ~45 minutes with 24-48 hours incubation
Day 5: ~45 minutes with 24-72 hours incubation
Day 6: ~45 minutes

The timeline to complete light induction with the Light-it Kit is:

Day 1: ~60 minutes to make LB agar petri dish and streak cells, with 24 hours incubation
Day 2: ~30 minutes to paint your living paintings with 24-48 hours incubation
Day 3: ~15 minutes to view results.
Day 3+: Repeat day 2/3 activities on any remaining petri dishes

Timeline to read *Fundamentals* is typically 3 hours.

Learning Hands-On: Manually turning on genes *in situ*

In situ means "in position of", and the hands-on exercises in this chapter involve first engineering *E. coli* bacteria with a DNA plasmid, and subsequently turning the desired genes on using an environmental cue while the system is in the bacteria. Because the cells will not be programmed to turn on the desired gene(s) right away, you'll notice that the engineered cells look like normal *E. coli* cells - they are white in color. Only once you enact the environmental cue will the cells begin the *Three Steps to Microfacturing* to create your desired end product.

In this chapter, you will use the skills you learned to genetically engineer cells. You can also complete some of the other exercises you've learned along the way, like lysis and extraction; you just need an Extract-it Kit! You would simply complete that exercise at the appropriate time, after you've induced and caused the cells to express the end product. Would the end product be any different than in cells that auto-induce expression? That sounds like a good experiment, you'll have to try it out to see. This is what genetic engineers and scientists do!

Exercise 1: Inducing a gene using a chemical

Step 1.   Complete the Induce-it Kit engineering exercise.

While in previous experiments your engineered cells would automatically express the desired trait, usually in the "late stationary phase" of growth (see Going Deeper 5-1), the plasmid you are now using is in the *mostly* 'off position' until you turn it on. We say *mostly* off since all gene control has some leakiness, which means that low/very low expression can happen even when the gene is "off". Upon successfully engineering your cells, they will be visually similar to the blank cells you streaked on the N.S. LB agar plates. Engineer the cells using **"Bag 1"** of the Induce-it Kit.

Step 2.   Culture your cells

Within the Induce-it Kit, there are two bags. "Bag 1", which you used in Step 1, included all of the ingredients needed for the initial engineering of bacteria with the chemically inducible plasmid. **"Bag 2"** contains the materials and ingredients to culture your cells, as well as induce expression of a gene using an inducer. Using the contents of "Bag 2", culture your cells on selective media as you usually would. You may use the streaking method you used with blank cells, or you can use the double streak method. Incubate your cells for 12-24 hours at 37 °C.

Step 3.   Add your inducer

A. Within the Induce-it Kit "Bag 2" you will find a **tube of inducer.** This inducer is in powder form. Because your kit was likely bumped around while in transit to you, the powder may have spread all throughout the tube.

To ensure you dissolve all of the powder, place the tube in your microcentrifuge and balance it. Spin the tubes at maximum speed for 10 seconds. This will pull all of the powder to the bottom of the tube.

B. The Inducer is in powder form so you will need to dissolve it in liquid to add it to your petri dish with the engineered cells. Using the **pipet** included in the kit, pipet all of the **'dissolving buffer'** into the tube of inducer powder. Pipet up (suck in) and down (squeeze out) 10 times to mix.

C. Using a **permanent marker**, make one or more marks, like an X, on the bottom of the plate in the area(s) where you want to put the inducer. This is where you will drop the inducer onto the plate.

D. Drop-by-drop, pipet all of the dissolved inducer onto your desired locations. The inducer will be at it's highest concentration at these points. However, it will dissolve outwards in a circular pattern through the LB agar. The higher the inducer concentration, the more the gene will be induced.

Inducer *Going Deeper* **7-1**

The inducer in this kit is a small molecule called isopropyl β-D-1-thiogalactopyranoside, or IPTG for short. IPTG mimics a sugar called lactose, the primary sugar in milk, which is used naturally by some cells (like the *E. coli* naturally in your intestines). When lactose is present, it will induce lactose metabolizing genes so that lactose can be used as an energy source.

The gene gets turned on by IPTG or lactose because IPTG binds to a protein called a repressor, which, in the absence of IPTG, binds to the promoter of the gene. Because the repressor binds to the promoter region, RNA polymerase cannot initiate transcription. When IPTG is added to the cells, it binds to the repressor protein and causes a slight change in shape of the repressor which lowers its ability to bind to the promoter region. The repressor 'falls off', allowing the RNA polymerase to initiate transcription (Figure 7-2).

Inducing cultures *Pro-tip*

In this exercise, you added the inducer after growing your engineered cells for 12-24 hours. Further, you induced only on part of the plate. This was for demonstration purposes, allowing you to see some cells exhibiting gene expression while others did not express the gene.

In many genetic engineering exercises, you would grow your engineered cells in liquid culture, and after ~12-16 hours of growth, add the inducer to the entire culture.

Exercise 2: Inducing a gene using temperature

Step 1.   Complete the Heat-it Kit engineering

Similar to Exercise 1, the Heat-it Kit plasmid you engineered remains mostly 'off position' until you decide to turn it on. The Heat-it Kit does not have as 'tight' switching as the Induce-it Kit. However, if you culture your cells at ~30 °C, the gene will stay mostly off. Complete the genetic engineering portion of your Heat-it Kit as you did in Chapter 4 using **"Bag 1"**. Continue on with **"Bag 2"** using the culturing technique you learned in Chapter 5. After streaking the engineered cells on your plates, incubate them at 30 °C for 24 hours. After 24 hours you should see lots of growth. However, the colonies should be mostly white in color. You can also opt to paint a picture with your engineered bacteria rather than streaking it!

Step 2.   Increase the temperature

In the case of temperature-based gene induction, there are no external factors that need to be prepared, other increasing the temperature of the environment. After your cells have grown and colonies are visible, you merely increase your DNA Playground incubator to 42 °C.

Increasing the temperature *Going Deeper* **7-2**

Increasing the temperature 'stresses' the cells. This causes a change in gene regulation throughout the cell, including the production of some different types of sigma factors. Recall back to Chapter 4 when you learned how sigma factors bind to the promoter of a gene and then to RNA polymerase. RNA polymerase can then begin transcribing the DNA into RNA. Some types of sigma factors are not created by the cell when the cell is in optimal growth conditions.

Optimal growth conditions include optimal temperature (37 °C), acidity/pH (~neutral), and other factors. By increasing the temperature to 42 °C, the cell 'realizes' that growth conditions are no longer optimal and begins creating sigma factors that will express genes to help the cell survive in the stressed environment.

For example, the promoter in the non-coding promoter region of the gene you are inducing is designed to bind a sigma factor called sigma E. When the cell enters 'stress mode' it begins producing sigma E which will bind to the promoter in your gene of interest facilitating transcription. In other words, this turns on or induces expression of your gene. You are now 'hacking' the survival mechanism of the cell to produce your proteins.

In some instances, if your cells grow into large colonies, or start dehydrating, this can activate the stress response and cause induction of your gene.

Exercise 3: Inducing a gene using light.

The Light-it Kit does not require any engineering, and is similar to the Canvas Kit exercise you did in Chapter 3. Complete the exercise as you did in the past by streaking a painting palette and using it to streak/paint pictures.

Step 1.   Streak cells

Just like in Exercise 1 and 2, the cells you are now using are in the 'off position' until you decide to turn them on. The cells will appear similar to the blank cells that you streak on the non-selective LB agar plates until you choose to activate the genes with light.

Culture your cells on one petri dish to make your painting palette. You will incubate them under a specific light to turn the color-producing gene on in the next step!

Step 2.   Turn on the light!

In the case of light-based gene induction, there are no external factors that need to be prepared, other than an environment with the appropriate wavelength of light. For the creation of your painting palette, you will streak the Light-it cells and incubate them under the Light-it Kit LED (light emitting diode). After your cells have grown and colonies are visible, you will use the painting palette to paint on your other petri dishes, and incubate at

37 °C with the light on and off by following the direction in the instructions manual.

For this system, the wavelength of light is critical. The CcaSR system needs 535 nm light to become activated. If your DNA Playground has the built-in LED module, turn it on. If you are using the standalone Light-it LED chamber, connect the battery by following the instruction manuals, then place it inside a 37 °C incubator. Note that accurate temperature is important for this induction to work.

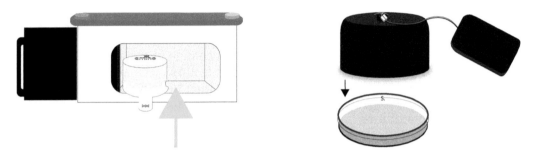

Figure 7-1. Use the built-in LED in your DNA Playground Large or the standalone Light-it LED chamber to shine light on your plates.

Light and bacteria *Going Deeper* **7-3**

Light is an essential sensory cue for many organisms. You use your eyes to sense light that is reflected off of the environment around you. Insects and animals use similar methods. Plants use light to understand growing seasons. Bacteria use light to understand the environment around them.

Phytochromes are proteins in plants and bacteria that are able to sense light radiation in many different wavelengths. The light may be visible, such as in the blue, green, or red spectrum, but may also be in the far red and infrared spectra (heat). In addition to sensing the light, phytochromes have the ability, through many different mechanisms, to activate and repress gene expression.

In the case of this exercise, the phytochrome CcaSR is from a genus of water-dwelling cyanobacterium, called *Synechocystis* that use it to regulate their circadian rhythm, and for phototaxis (movement based on light).

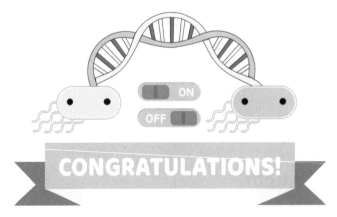

CONGRATULATIONS!

on completing your sixth set of experiments!

Fundamentals: Diving deeper into genetic 'switches'

There are hundreds of known and countless unknown genetic "switches" that enable cells to turn gene expression on or off based on the environmental conditions outside of the cell, or the physiological state inside of the cells. In the exercises within this chapter, you explored three different classes of gene regulation - chemical, temperature, and light. There are many more gene regulation systems, including many that are unknown to the authors!

In this section, we are going to explore more in-depth how each of the gene regulation pathways used in the hands-on exercises works, and relate their function back to the original operation that occurs in the cells.

Turn on genes with chemicals

Chemicals can be broadly defined as any molecule. Genes can be turned on or off by many different molecules ranging from molecular oxygen gas, to sugars, and even heavy metals like arsenic.

In Exercise 1, the genetic system you used was inspired by a genetic regulatory network called the lac operon. The lac operon was discovered nearly 100 years ago in *E. coli* bacteria and it unveiled for the first time the sophistication of genetic circuits in cells.

However, rather than look at how the entire lac operon genetic system functions, we will focus on one part which you used in the hands-on. The critical molecules that are involved in the chemical induction exercise include:

Gene of Interest:

1. **The coding region** of the gene contains the DNA sequence for colored pigment. When transcription and translation occur, colored proteins are created. In Chapter 5 you learned that in translation, the coding region has a start codon at the 5' end and a stop codon at the 3' end.

2. **The non-coding region** of the gene is where the magic happens in this exercise. You are familiar with promoters (Chapter 4) and the ribosomal binding site (Chapter 5). They are important for RNA polymerase and ribosome binding, which initiates transcription and translation. In this

genetic circuit, there is also a short sequence within the promoter called an operator. Similar to how the promoter and RBS are able to bind to proteins such as RNA polymerase and the ribosome, the operator is a short segment of DNA that is able to bind to other proteins. These proteins can either activate gene expression with an 'activator' (similar to a sigma factor) or repress gene expression with a 'repressor'. The repressor represses gene expression by preventing RNA polymerase from binding to DNA or initiating transcription. In the exercise you completed, there is a repressor called the lac repressor which is able to bind to the operator tightly and stay there (Figure 7-2).

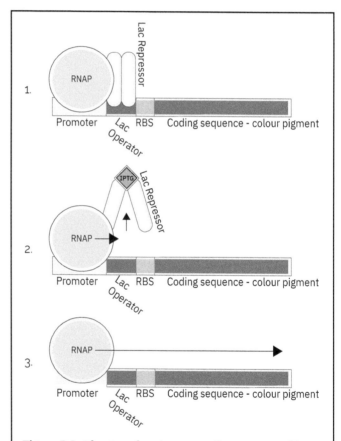

Figure 7-2. After transforming your cells, your gene of interest is: 1) in the 'off' state because lac repressor is bound to the lac operator; 2) the added IPTG inducer binds to lac repressor which changes its shape so that it can no longer bind to the lac operator and block RNA polymerase; 3) with the lac repressor removed, the RNA polymerase initiates transcription.

Lac repressor: Lac repressor is a protein that is also expressed in the plasmid that you engineered into your K12 *E. coli*. Lac repressor is capable of binding tightly to the lac operator. When bound, the lac repressor acts like a roadblock, preventing RNA polymerase from initiating transcription, meaning that gene expression cannot occur. (Figure 7-2, 7-3).

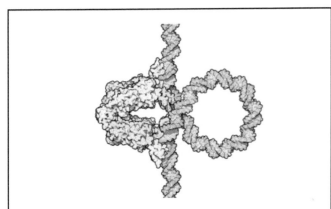

Figure 7-3. Lac repressor (green) is a protein with 'two feet' that can bind tightly to the lac operator which can be embedded within the promoter of a gene. This action prevents the RNA polymerase from proceeding with transcription. Source: D. Goodsell, S. Dutta, C. Zardecki, M. Voigt, H. Berman, S. Burley. (2015) The RCSB PDB "Molecule of the Month": Inspiring a Molecular View of Biology. PLoS Biol 13: e1002140.

Inducer: IPTG is a small chemical molecule that is used as an inducer, meaning it can cause induction (activation) of gene expression. IPTG is a molecule synthesized in laboratories that is similar to a natural molecule called lactose (Figure 7-4).

The natural inducer, lactose, is a sugar that activates gene expression by binding to lac repressor. In the natural lac operon in cells, the protein enzyme created upon activation of gene expression is called beta-galactosidase. Beta-galactosidase is a protein that ultimately cuts the lactose sugar in half to become glucose and galactose (Figure 7-5).

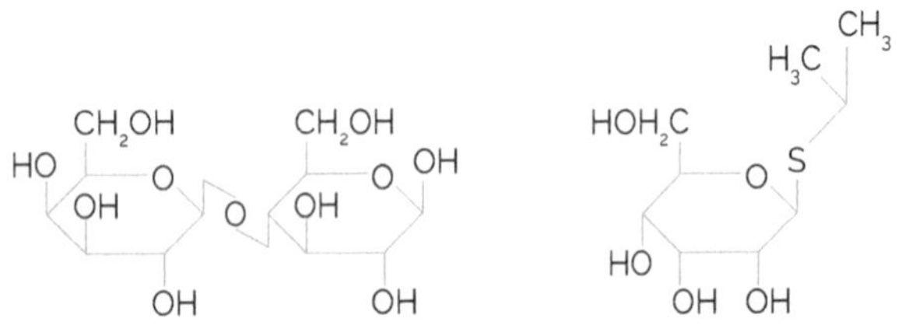

Lactose

Isopropyl-thiogalactoside (IPTG)

Figure 7-4. Lactose is the natural sugar inducer that binds with lac repressor to remove it from blocking transcription. IPTG is a chemical synthesized by scientists that is much more stable than the natural inducer and can be used in place of lactose

In other words, the lactose molecule activates gene expression to create an enzyme that destroys it! When it destroys the lactose, the galactose and glucose become a carbon energy source for the cell. The practical reason for this is that the cell does not want to create enzymes when they don't need them. Why should the cells always create beta-galactosidase if there is no lactose present? Only when lactose is present will it create the enzymes to break the lactose into a usable food source. Recall in the Blue-it Kit (Chapter 6), you engineered cells to create beta-galactosidase, the same enzyme that originates from the lac operon!

Now that you know the natural function of the lac repressor, lac operator, and lactose inducer, you may realize that the lactose inducer isn't really the ideal inducer for genetic engineers. This is for two reasons:

1. Lactose can be consumed by bacteria as an energy source. This means that when you add the inducer to the system, the cells will slowly consume lactose until none is left and induction stops.

2. Because lactose is an energy source for cells, by adding it, you are changing the cells' metabolism. This can create variation in how the cells operate and the results of your genetic engineering experiment might be more variable.

IPTG was created by scientists to overcome both of these issues. IPTG is not consumed by the cell, and therefore it cannot directly influence the energy systems of the cell. Second, because IPTG cannot be consumed by the cell, the number of IPTG molecules that you add to the system remains constant. This enables much better control of the induction of genes.

To summarize the activity that occurs when you induced gene expression with IPTG:

1. After you genetically engineer the *E. coli* with the plasmid, the gene of interest is in the 'off state'. This is because lac repressor, which is constitutively expressed (automatically turns on) from another gene in the same plasmid, binds to the lac operator of your gene of interest, blocking RNA polymerase from initiating transcription.

2. Upon adding IPTG inducer to the petri dish, the IPTG crosses the cell membrane into the cell cytoplasm and completes the Four B's. When it interacts with lac repressor, it is able to bind, and this changes the shape of the lac repressor, causing it to detach from the lac operator.

3. With the lac repressor removed from the gene of interest, the RNA polymerase can initiate transcription.

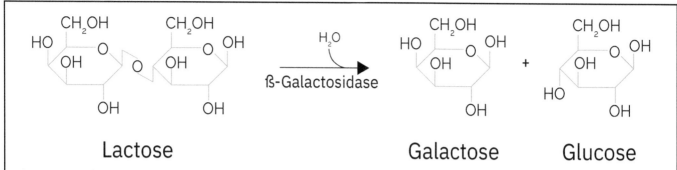

Figure 7-5. In the natural lac operon, upon removing the lac repressor, the enzyme beta-galactosidase (beta-gal) is expressed. Beta-gal has the function of cutting lactose between its sugar rings, resulting in galactose and glucose.

Turn on genes with temperature

In Chapter 4, you learned that to start transcription you need a promoter that sigma factors bind to, which then "recruit" or bind to RNA polymerase to initiate transcription. *E. coli* cells have many different sigma factors that are expressed in the cell under many different conditions and are always working in the background to make sure that essential cellular stuff, such as tRNAs and polymerases, are being created. For example, under normal cell operation, where the cells are grown in a favorable environment at 37 °C, with plenty of food, a sigma factor called sigma 70 (σ70) will be automatically expressed by genes in the cell. σ70 is responsible for causing expression of many "housekeeping" genes that enable overall cell survival and growth.

In the plasmids you have been using for your genetic engineering exercises, many of the constitutive promoters that caused expression of the genes of interest, such as color pigment genes, are recognized and bound to by σ70.

If the cell encounters starvation or unfavorable environmental conditions such as heat, other sigma factors are expressed. These sigma factors are able to cause expression of a different set of genes that are typically off, to aid with helping cells to survive. Incubating *E. coli* bacteria at temperatures of 42 °C to 50 °C is very stressful to the cells and causes the expression of such sigma factors. σE is a sigma factor that is related to *E. coli* temperature stress response.

By incubating the cells at 42°C and above, a significant change happens in the cells, and they begin to use a sigma E (σE) to cause expression of genes. The plasmid you used in this exercise uses σE promoter so that when the temperature increases in the cells' environment, σE can bump into and activate expression of your color gene. (Figure 7-6).

In the case of the exercise you completed, it is as simple as using the cells natural response to danger or stress to your advantage. Unfortunately, *E. coli* cells do not enjoy high temperatures, nor can they survive for extended periods of time, so using this strategy as a long-term tool is not ideal. *E. coli* cells can survive at up to 46 °C for prolonged periods of time, but anything hotter will lead to their demise.

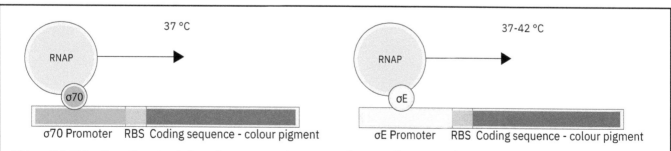

Figure 7-6. Using the cells natural sigma factor expression to control a gene of interest. Left: Under favorable cell operations, sigma 70 recruits RNA polymerase to initiate transcription of your gene of interest; Right: Under stress such as heat, the cell creates sigma E, which can be used to initiate transcription of your gene of interest.

Turn on genes with light

Light is one of the most important environmental cues for cells and all living organisms. Being able to see physical objects in your environment at a distance is key to survival for many organisms. Many organisms use light as an energy source. For example, plants and bacteria photosynthesize. While these organisms are harnessing light as an energy source, they can also learn about and respond to their environment. They can measure the amount of certain colors of light.

Responding to the environment, whether you are a human or a bacteria, involves light photons hitting molecules or macromolecules either on or inside the cells. Those molecules may harness energy from the light photons to change shape or to cause a chemical reaction to happen. In the case of the hands-on exercise where you induced gene expression using light, there are several different cellular systems at play. The plasmid that your *E. coli* were engineered with includes several different players. The following section is an in-depth discussion of the light induction system.

CcaS: is a protein that is constitutively expressed in the cell after engineering the cell with a plasmid. CcaS has two parts, called domains, that have two different functions. One is the light receptor domain, which is able to absorb green light of ~532 nm. The second domain is called a kinase (k-eye-naise). A kinase has the function of adding phosphate (PO_4) molecules to other molecules in the cell, such as proteins. As you have seen throughout this book, phosphate is highly negatively charged. By adding a phosphate to a protein, the negative charge will change the shape of the protein through ionic bonding and hydrogen bonding. When the shape of a protein changes, it can be activated to complete a chemical reaction, or inactivated to stop completing a chemical reaction.

When the light receptor domain absorbs light, the kinase domain of the protein becomes active and is able to phosphorylate (add a phosphate to) another protein in the cell. The protein that becomes phosphorylated in this system is called CcaR.

CcaR: is a protein constitutively expressed in the cell upon transforming the cell with the plasmid. CcaR has two domains as well. One domain is able to specifically bind to CcaS so that it can become phosphorylated. When CcaR becomes phosphorylated, its shape changes so that the other domain can bind specifically to a promoter called pCpcG2-172. CcaR can act like a sigma factor and cause RNA polymerase to bind and initiate transcription. In the plasmid you used, the light activation of CcaR causes the expression of a second sigma factor called CCG.

Sigma CCG: In the plasmid that was pre-engineered into *E.coli* cells, the pCpcG2-172 promoter is placed in front of a coding sequence for another transcription factor that will activate the expression of your gene of interest. In this case, the protein CCG is able to bind the pCCG promoter, which causes transcription of your gene of interest.

Core T7 polymerase: T7 polymerase is a very popular polymerase that genetic engineers use to selectively transcribe coding sequences. T7 polymerase is an RNA polymerase that comes from a bacterial virus called a phage. It can transcribe from a specific promoter called a T7 promoter because it has a 'built-in sigma factor' that will bind to the T7 promoter. This means that no *E. coli* bacteria will naturally create T7 polymerase, and nor will the cells be able to transcribe from a T7 promoter. In the case of the plasmid you used in this exercise, the T7 polymerase has been "cut in half" so that only the transcribing part of the T7 polymerase is constitutively created by the cell. The genetically engineered sigma factor called sigmaCCG is expressed due to CcaR binding to pCpcG2-172. CcaR binding to pCpcG2-172 happens when the right light is present. Sigma factor CCG binds to the CCG promoter, pCCG, which is at the start of your color-producing gene. The T7 Core polymerase binds to sigma factor CCG and transcribes your the coding sequence for your color pigment.

This is a pretty complex genetic pathway, but it will illustrate the sophistication of genetic regulation in cells and that there is a lot of potential for innovating in genetic engineering (Figure 7-7)! To summarize the light activation system:

1. As in other plasmids you've used, the antibiotic selection gene is designed to automatically create antibiotic resistance so you can select for your engineered bacteria.

2. Rather than using the cell's natural RNA polymerase, you are using a unique polymerase called T7 polymerase that will be specific to the genes you want to express in the plasmids in the pre-engineered bacteria. Automatic expression is used so that the RNA polymerase is 'ready and available' for transcription of the color-producing gene in the plasmid.

3-4. Light sensor proteins CCaS and CCaR are also automatically expressed so that they can be ready for when the right light is present.

5. When the right light is present and hits the CCaS protein, the CCaS becomes active and can add a phosphate molecule (phosphorylation) to the CCaR protein. This allows CCaR to bind to the pCpcG2-172 promoter and activate gene expression.

6. The phosphorylated CCaR protein binds to a promoter called pCpcG2-172 which activates expression of a sigma factor called CCG that is only able to interact with the T7 RNA polymerase (Step 2) and another promoter that is the non-coding region of your color-producing gene.

7. Sigma CCG is expressed and completes the Four B's to ultimately interacts with the pCCG promoter upstream of the pigment coding sequence. Upon binding to the pCCG promoter, sigmaCCG also binds to the T7 RNA polymerase that was automatically created in Step 2. This leads to transcription of the pigment coding sequence, which is followed by translation and folding of the pigment protein. The bacteria change color!

8. Sigma CCG and T7 RNA polymerase start the *Three Steps to Microfacturing*, and a pigment protein is created.

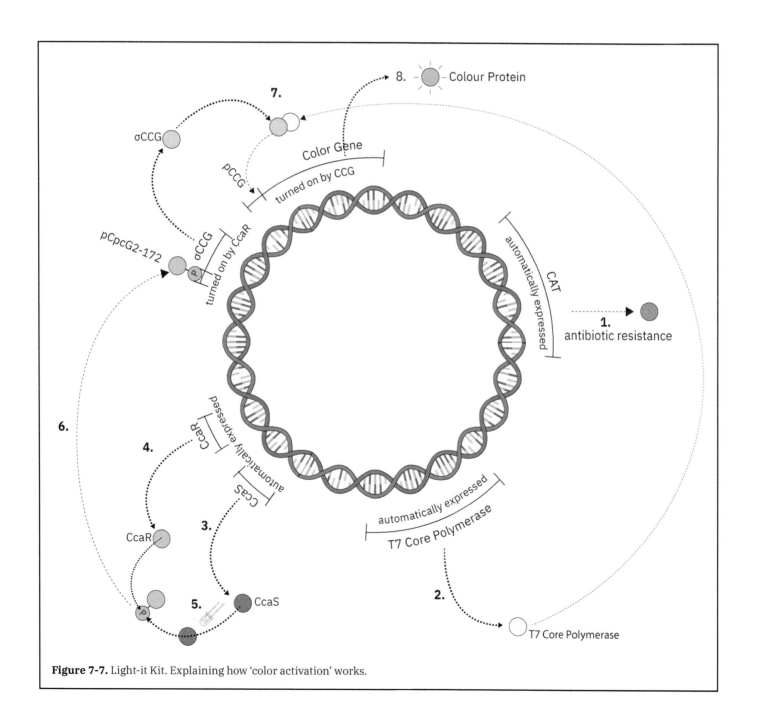

Figure 7-7. Light-it Kit. Explaining how 'color activation' works.

Summary

In this chapter, you engineered K12 *E. coli* with DNA plasmids with genetic regulation that can be controlled. While some of the plasmids' genes were similar to those used in previous experiments, like the selection marker genes that automatically turned on, each of the plasmids had a gene of interest that was controllable by a different class of environmental cue - chemical, temperature, and light.

A common theme is that some proteins can either activate (turn on) or repress (stop) transcription. A sigma factor that binds to the promoter region recruits RNA polymerase (or T7 core polymerase) to cause transcription activation to occur. Conversely, a protein repressor bound to the operator prevents transcription until it is removed from the operator. In fact, during activation, sigma factors (also called transcription factors) needed to be either created naturally by the cellular genes, by the genes in the plasmid you transformed the cell with, or through chemical modification such as phosphorylation. In each of these instances, the proteins then bound to the promoter region to recruit polymerase. During repression, a protein was 'roadblocking' the RNA polymerase, and an inducer molecule was added to the cells, which bound to the repressor, changed its shape, causing it to no longer bind to the DNA promoter. This allowed RNA polymerase to initiate transcription.

You are now well versed in growing cells, genetic engineering, extracting proteins, enzymatic reactions, and understanding genetic regulations of DNA plasmids. Congratulations, Genetic Engineering Hero!

Review Questions

Hands-on Exercise

1. What does IPTG stand for and what does it do?

2. List two examples of how microorganisms use light to manage their lives.

3. How are "optimal growth conditions" and "stressful growth conditions" different?

Fundamentals

1. When turning on the Exercise 1 genetic system with chemicals, what happens that makes RNA polymerase able to begin transcription?

2. Why is IPTG used in the Induce-it Kit experiment rather than lactose?

3. What is the non-coding region called that lac repressor binds to?

4. During the heat induction experiment, explain how increasing the heat makes it possible for your gene to be expressed.

5. What is the difference between constitutive expression and inducible expression?

6. What is T7 polymerase? Where is it from? How is it used in the light induction system?

7. How is phosphate used during the light induction system?

8. List three different molecules/systems that phosphate is used in within *E. coli* cells.

The Chemical Elements of *E. coli* Biochemistry

1 1.008 **H** Hydrogen	

3 6.941 **Li** Lithium	**4** 9.012 **Be** Beryllium
11 22.990 **Na** Sodium	**12** 24.305 **Mg** Magnesium

19 30.098 **K** Potassium	**20** 40.078 **Ca** Calcium	**21** 44.956 **Sc** Scandium	**22** 47.88 **Ti** Titanium	**23** 50.942 **V** Vanadium	**24** 51.996 **Cr** Chromium	**25** 54.938 **Mn** Manganese	**26** 55.847 **Fe** Iron	**27** 58.933 **Co** Cobalt
37 85.468 **Rb** Rubidium	**38** 87.62 **Sr** Strontium	**39** 88.906 **Y** Yttrium	**40** 91.224 **Zr** Zirconium	**41** 92.906 **Nb** Niobium	**42** 95.95 **Mo** Molybdenum	**43** 98.907 **Tc** Technetium	**44** 101.07 **Ru** Ruthenium	**45** 102.91 **Rh** Rhodium

Nucleotides (in DNA and RNA)

Lipids (in Fats and Lipids)

Sugars and Starches (in Carbohydrates)

Amino Acids (in Proteins)

Salts/Minerals (Frequently Used in Cells)

Salts/Minerals (Rarely Used in Cells)

1 H Hydrogen	2 He Helium

5 B Boron	6 C Carbon	7 N Nitrogen	8 O Oxygen	9 F Fluorine	10 Ne Neon
13 Al Aluminium	14 Si Silicon	15 P Phosphorous	16 S Sulfur	17 Cl Chlorine	18 Ar Argon

28 Ni Nickel	29 Cu Copper	30 Zn Zinc	31 Ga Gallium	32 Ge Germanium	33 As Arsenic	34 Se Selenium	35 Br Bromine	36 Kr Krypton
46 Pd Palladium	47 Ag Silver	48 Cd Cadnium	49 In Indium	50 Sn Tin	51 Sb Antimony	52 Te Tellerium	53 I Iodine	54 Xe Xenon

 CHOPNS

Milton Keynes UK
Ingram Content Group UK Ltd.
UKHW051923310823
427861UK00006B/50